An Identifica
Common Insects,

Backyard Bugs

Jaret C. Daniels

Adventure Publications
Cambridge, Minnesota

Acknowledgments

I would like to thank my loving wife, Stephanie, for her unending patience, sense of humor, and support with the countless number of butterflies, caterpillars, and other assorted insects that always happen to find their way into our home as part of projects like these. I would also like to thank my parents for encouraging my early interest in the natural world. It has resulted in a continuously rewarding and always surprising career.

Cover and book design by Lora Westberg

Edited by Brett Ortler

Anatomy illustration by Julie Martinez

Photo credits:
Cover photos: Shutterstock
The following photos are copyright of their respective photographers. **Scott Bauer, USDA Agricultural Research Service, Bugwood.org:** 168 (main) **Joseph Berger:** 202 (inset) **Joseph Berger, Bugwood.org:** 202 (main) **Ray Brunn:** 109 (both) **Fitz Clarke, Savannah, GA:** 82 (inset) **John Flannery:** 121 (inset) **Bill Keim:** 29 (inset) **Warren P. Lynn:** 30 (main) **Jim McCulloch:** 96 (main) **Copyright 2010 Cheryl Moorehead:** 43 (main) **Tom Murray:** 104 (inset) **Brett Ortler:** 16 **Richard Payne Jr.:** 30 (inset) **Robert Peal:** 194 (inset) **Lawrence E. Reeves:** 61 (main), 68 (main), 140 (inset) 218 **Alan Schmierer:** 106 (inset) **Yellowstone National Park, National Park Service. J. Schmidt:** 195 (inset)

The following (unaltered) images are licensed according to a Creative Commons 2.0 License (https://creativecommons.org/licenses/by/2.0/):

Page 54 (main), "Whirligig beetle, Gyrinidae," by Andy Reago & Chrissy McClarren. Original available at www.flickr.com/photos/wildreturn/18388506375/

Page 104 (main), "Black and Yellow Lichen Moth" by Andy Reago & Chrissy McClarren. Original available at www.flickr.com/photos/wildreturn/9887606146/

All other photos by Jaret C. Daniels or Shutterstock

10 9 8 7 6 5 4 3 2

Copyright 2017 by Jaret C. Daniels
Published by Adventure Publications
An imprint of AdventureKEEN
820 Cleveland Street South
Cambridge, Minnesota 55008
(800) 678-7006
www.adventurepublications.net
All rights reserved
Printed in China
ISBN: 978-1-59193-685-5; eISBN: 978-1-59193-686-2

How to Use This Book

This book is organized by where you're likely to encounter specific insects. There are seven main sections—At Lights, In or Near Water, In the Air, On Flowers, On Structures, On the Ground, and On Plants. Of course, these aren't hard-and-fast rules, but they are a good place to start identifying the insects you find.

Within each section, insects are organized by their order—a scientific grouping of similar insects. For example, butterflies and moths all belong to an order—Lepidoptera. Many of these orders are relatively familiar—and we've provided example species to help guide you—but some are likely less familiar. To help you, each account includes full-color photos of each type of insect; if you don't recognize what kind of bug you've found, follow these steps.

1 Look in the section where you found the animal.

2 If you know what kind of critter it is (a butterfly or moth), look for that grouping.

3 If you don't know it, flip through and look at the photos to find bugs with a similar shape to find its order.

4 Within each order, the insects and other critters are organized by average size, from smallest to largest.

Butterflies and Moths (Order Lepidoptera)

Imperial Moth
Size: Wingspan 3.5–6.8 inches
ID Tips: Large; somewhat elongated yellow wings with a varying degree of purplish-brown markings
Range: The eastern United States

With females often having a wingspan of more than six inches, this impressive insect is one of the largest moths in North America. Its elongated yellow wings have varying amounts of purplish-brown markings and tend to resemble fallen leaves. This wing pattern likely provides effective camouflage, helping moths resting during the daytime blend unnoticed into the background vegetation. While adults are often attracted to artificial lights, males are more frequently encountered than females. The stout larvae may be either green or brown and are covered with fine hairs and have four short knobby horns behind the head. They feed on a wide range of trees, including oak, pine, maple, and hickory. When fully grown—and approaching some five inches in length—the larvae crawl down and wander extensively. Once a suitable location is found, they burrow into the soil and pupate underground.

Believe It or Not: These sizeable caterpillars produce extensive amounts of large, barrel-shaped fecal pellets called frass. These conspicuous droppings can often be spotted on the ground beneath an occupied tree.

Note: While this book focuses mostly on insects, we've also included a number of other "creepy crawly" animals, such as spiders, centipedes and millipedes, earthworms, and slugs. Such animals are often lumped together with insects, and while they aren't technically insects, they are common and interesting enough to merit including.

Table of Contents

How to Use This Book 3

Insect Anatomy 6
   The Head 7
   The Thorax 7
   The Abdomen 7

Stages of Development 8

What You Might Find 8

Where to Look for Insects 14

Beware of Bites and Stings 24

Bugs Found at Lights
   True Bugs
     (Order Hemiptera) 26
   Megaloptera (Alderflies,
     Dobsonflies, and Fishflies) 27
   Butterflies and Moths
     (Order Lepidoptera) 28
   Beetles (Order Coleoptera) 41
   True Flies (Diptera) 45
   Antlions, Lacewings, and Mantidflies
     (Order Neuroptera). 46
   Spiders (Order Araneae) 48
   Mayflies (Order Ephemeroptera). . 49
   Caddisflies (Order Trichoptera). . . 50
   Stoneflies (Order Plecoptera) 51
   Grasshoppers, Crickets, Katydids,
     Locusts, and others
     (Order Orthoptera) 52

Bugs Found in or Near Water
   Beetles (Order Coleoptera) 54
   Flies (Order Diptera) 55
   True Bugs (Order Hemiptera) 56
   Dragonflies and Damselflies
     (Order Odonata) 59

Bugs Seen in the Air
   True Flies (Order Diptera) 68
   Beetles (Order Coleoptera) 70

Bugs Found on Flowers
   True Flies (Order Diptera) 72
   Beetles (Order Coleoptera) 75
   True Bugs (Aphids, Cicadas,
     and others) 81
   Wasps, Bees, Ants, and Sawflies
     (Order Hymenoptera) 82
   Spiders (Order Araneae) 92
   Butterflies and Moths
     (Order Lepidoptera) 94

Bugs Found on Structures
   Bees, Wasps, and Ants
     (Order Hymenoptera) 134
   True Flies (Order Diptera) 136
   Spiders (Order Araneae) 137

Bugs Found on the Ground
   Sowbugs, Pillbugs, and Woodlice
     (Order Isopoda) 138
   Beetles (Order Coleoptera) 139

Spiders (Order Araneae) 145

Bees, Wasps, and Ants
(Order Hymenoptera) 146

Butterflies and Moths
(Order Lepidoptera) 150

Millipedes (Order Julida) 152

Cockroaches and Termites
(Order Blattodea) 153

Earthworms
(Order Megadrilacea) 155

Vinegaroons (Order Uropygi) . . . 156

Lacewings, Mantidflies, Antlions
(Order Neuroptera) 157

Snails (Order Pulmonata) 158

Slugs (Order Soleolifera) 159

Earwigs (Order Dermaptera) . . . 160

Harvestmen and Daddy Longlegs
(Order Opiliones) 161

Centipedes (Class Chilopoda) . . . 162

Grasshoppers, Crickets, and Katydids
(Order Orthoptera) 163

On Vegetation

Aphids, Cicadas, and Others
(Order Hemiptera) 164

Beetles (Order Coleoptera) 179

Butterflies and Moths
(Order Lepidoptera) 186

Grasshoppers, Crickets, Locusts, and
Others (Order Orthoptera) . . . 202

Spiders (Order Araneae) 209

True Flies (Diptera) 211

Dragonflies and Damselflies
(Odonata) 213

Mantises (Order Mantodea) 214

Walking Sticks (Order Phasmida) 215

Fun and Family-Friendly Bug Activities 216

Easy

Netting Insects 216

Hunting for Wolf Spiders
with a Flashlight 217

Butterfly Watching 217

Attracting Insects
with a Black Light 217

More Advanced Projects

Moth Baiting 219

Native Bee Nest Box 219

Planting a Pollinator Garden . . 221

Rearing Caterpillars 222

Pitfall Trapping 223

About the Author 224

With nearly 1 million species recorded, insects are the most diverse group of organisms on the planet. They account for approximately 75 percent of all described animal species, and the vast majority of the species on Earth that are yet to be identified are likely insects and other arthropods. In fact, scientists conservatively estimate that the total number of insect species could exceed 8 million when all are eventually discovered. While the majority of that amazing diversity exists in the tropics, there are well over 150,000 insect species found within the United States and Canada alone and many more if other arthropods, the larger group to which insects, spiders, centipedes, and scorpions all belong.

Insect Anatomy

Insects share several common characteristics. Unlike mammals, birds, reptiles, and amphibians, insects lack an internal skeleton. Instead, insects have a hard exoskeleton on the outside of their bodies; this provides both protection and support. Their body is divided into three distinct regions: the head, thorax, and abdomen.

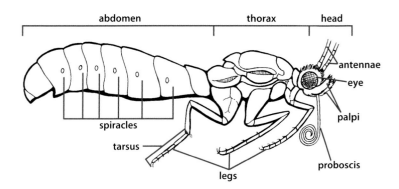

The Head

An insect's head has two rather prominent compound eyes, two antennae, and mouthparts. The rounded compound eyes are composed of hundreds of tiny individually lensed eyes. Together, they render a single, somewhat pixilated image and afford insects rather good vision, especially for both distance and motion. Above the eyes are two antennae. They bear sensory structures that help with orientation, smell, and taste. The head also bears mouthparts, which vary considerably across insect groups; some insects, such as grasshoppers, beetles, and caterpillars, have chewing mouthparts. Others, such as true bugs and mosquitoes, have piercing-sucking mouthparts, but there are many other options, and they include mouthparts adapted for sponging (flies), rasping-sucking (thrips, biting flies), and siphoning (butterflies and moths). Some insects have reduced or vestigial mouthparts or may even lack them altogether.

The Thorax

The thorax is an insect's second body section. It bears the appendages and muscles that enable an insect to move. All insects have three pairs of jointed legs, one pair on each segment. Many insects also have one or two pairs of wings. Besides flight, wings may serve a variety of other functions, including assisting in thermoregulation, sex recognition, sound production, camouflage, mimicry, and self-defense.

The Abdomen

The last section of an insect's body is its abdomen, which contains the reproductive, digestive, and excretory systems along with a series of small lateral holes, called spiracles, that enable air exchange. In female insects, the tip of the abdomen may have an added structure called an ovipositor, which is used to insert or place eggs. In some insects, such as bees and wasps, the ovipositor is modified into a stinger that can be used for self-defense.

Stages of Development

All insects pass through a series of developmental stages as they grow. This transition is known as metamorphosis. Most insects, including butterflies and moths, flies, beetles, and bees and wasps, undergo a complete metamorphosis consisting of four developmental stages: egg, larva, pupa, and adult. The immature stages look much different than the adults, eat entirely different foods, and live in separate environments. Butterflies and moths are good examples. The wormlike larvae have chewing mouthparts; they feed primarily on plant material, live on plants, and have a mostly sedentary lifestyle. The adults have siphoning mouthparts, feed primarily on flower nectar or other liquid resources, and are highly mobile thanks to their wings.

Other insect groups, including grasshoppers, dragonflies and damselflies, true bugs, and cockroaches, undergo incomplete metamorphosis. This process consists of three developmental stages: egg, nymph, and adult. The young nymphs often closely resemble the adults in appearance, although they are smaller in size. They also may share the same environment and food resources that adults frequent, and they often behave similarly. True bugs are a good example. The nymphs are essentially smaller versions of the adults. They have piercing-sucking mouthparts, feed on plant sap, live primarily on plant material, and typically move by walking. The resulting adults differ in that they are larger, reproductively active, and have wings, enabling flight.

What You Might Find

Arthropods

Arthropods belong to the phylum Arthropoda and represent the largest group in the animal kingdom. Insects, spiders, millipedes, centipedes, and crustaceans are all arthropods; together, the arthropods are the most successful group of organisms on the planet! Arthropods share many common features, including a protective external skeleton made of chitin, a segmented body, and paired and jointed appendages. Insects are often the most noticeable and commonly encountered terrestrial group. They can be divided into more than 30 different orders (subgroups). The members of each have certain basic characteristics and behaviors that can be particularly useful for identification. The following include some of the most charismatic, distinctive, and commonly encountered orders covered in this book.

Coleoptera (Beetles)

This is the largest and most diverse group of animals on the planet. Beetles represent about 40 percent of all known insects with over 350,000 described species. Adults tend to be rather large, robust, and conspicuous organisms with a hard exoskeleton and two pairs of wings. The first pair is modified into protective covers, called elytra, which cover the larger, membranous wings beneath. Beetles occur in both terrestrial and aquatic freshwater environments. They have chewing mouthparts; some feed as predators on a variety of other invertebrates, others are herbivores and consume various plant parts, and still others are scavengers on animal dung, carrion, or decaying plant material. Beetles undergo complete metamorphosis. Many beetles are also attracted to artificial lights at night.

Lepidoptera (Butterflies and Moths)

Butterflies and moths are some of the most well-known and charismatic of all insects. Adults tend to be rather large and showy organisms with two pairs of transparent wings that are covered with numerous tiny scales. They have two large compound eyes, two elongated antennae, and siphoning mouthparts (although in some species they are significantly reduced and nonfunctioning), which enable them to drink flower nectar or other fluids. Butterflies are active by day, whereas most moths are nocturnal, although some species are seen during the day. Butterflies and moths undergo complete metamorphosis. Their larvae, known as caterpillars, are primarily plant feeders and have chewing mouthparts.

Hymenoptera (Bees, Wasps, and Ants)

This is a large and diverse group of insects known for their complex social systems that even include division of labor. Adults typically have two pairs of transparent wings, although some groups or individuals may be wingless. Most have chewing mouthparts with the exception of bees, which have a tongue for feeding on flower nectar or other fluids. The majority of adults also have a noticeable constriction between the abdomen and thorax that resembles a narrow waist. Many adults feed on nectar and are common flower visitors. Ants are typically predators, omnivores, or scavengers. All bees, wasps, and ants undergo complete metamorphosis. Their larvae feed on a variety of resources, including prey or pollen provisioned by the adults and plant material. Some are parasites of other insects.

Diptera (Flies)

Flies are a large and cosmopolitan group of insects that get their name for their most obvious behavior—flying. The rear wings are modified and reduced to small, club-shaped structures called halters, which help stabilize the insect during flight. Adults have well-developed compound eyes, short antennae, and diverse mouthparts designed for piercing, sucking, or sponging up liquid foods. Flies undergo complete metamorphosis. The larvae lack legs and live in terrestrial, freshwater, or moist environments where they feed on decaying plant or animal material or are predators or parasites of other animals.

Hemiptera (True Bugs)

This is a large and highly diverse group of insects, especially in size and appearance. Adults have two pairs of transparent wings or forewings that are partially thickened at the base. They have piercing or sucking mouthparts, with most species feeding on plant juices, although some are predatory. True bugs undergo incomplete metamorphosis with immature specimens (called nymphs) closely resembling adults. They are predominantly terrestrial, although a few groups occur in freshwater habitats.

Orthoptera (Grasshoppers, Crickets, Katydids, and Others)

This is a group of larger, robust insects found entirely on land. Adults have two pairs of wings; the first pair is narrow, hardened and leathery and covers a larger, membranous pair below. Both pairs of wings are held over the back while at rest. The hind legs are enlarged and modified for jumping. These insects undergo incomplete metamorphosis with the immature ones (called nymphs) closely resembling adults. Both have chewing mouthparts and feed primarily on plant material.

Odonata (Dragonflies and Damselflies)

This is a small but diverse group of insects. Adults have two pairs of large, transparent wings with extensive veinlike features on them; large compound eyes; and long, slender abdomens. Many adults are brightly colored and showy. They are active, mobile predators with chewing mouthparts. Dragonflies and damselflies undergo incomplete metamorphosis. The immature ones, known as naiads, occur in freshwater systems where they feed on other aquatic organisms.

Blattodea (Cockroaches and Termites)

This is a small and primarily ancient group of insects. Cockroaches are characterized by oval and somewhat flattened bodies, very long antennae, and two pairs of membranous wings. They have chewing mouthparts and are generally considered omnivores.

Neuroptera (Lacewings and Antlions)
This is a group of delicate-looking insects with two pairs of heavily veined, transparent wings. Adults have chewing mouthparts and feed on other insects. The larvae are active predators and have modified jaws to capture prey and suck out the internal fluids. They undergo complete metamorphosis. The adults are weak fliers and are often attracted to artificial lights.

Mantodea (Mantises)
Mantises are large and very charismatic insects. They have an elongated body with a distinctive triangular head, large compound eyes, chewing mouthparts, two pairs of wings, and enlarged front legs that are modified with spines and enable them to capture prey. Both juveniles and adults are highly camouflaged, sit-and-wait predators. They undergo incomplete metamorphosis. Adults are often attracted to artificial lights.

Phasmida (Walking Sticks)
This is a group of large, primarily tropical insects with few species in North America. They have chewing mouthparts and very elongated bodies, legs, and antennae that help them resemble sticks or other vegetation for protection from predators. Most species in our area lack wings. They undergo incomplete metamorphosis.

Dermaptera (Earwigs)
Earwigs are a distinctive and somewhat primitive-looking group of insects. They have elongated, flattened bodies; chewing mouthparts; somewhat elongated antennae; and a distinctive pair of pincer-like features called cerci on the tip of their abdomen. They have two pairs of wings, with the front pair noticeably short and leathery. They undergo incomplete metamorphosis.

Ephemeroptera (Mayflies)
Mayflies are a small group of distinctive and delicate-looking insects. They have elongated bodies, two pairs of transparent wings with the forewings much longer than the hind wings, short antennae, and three elongated filaments off the abdomen. The adults are extremely short-lived and thus have no functional mouthparts. Immature specimens are aquatic. They undergo incomplete metamorphosis. Adults are regularly attracted to artificial lights.

Megaloptera (Alderflies, Dobsonflies, and Fishflies)
This is a group of medium- to large-size, primitive-looking insects. They have elongated, soft bodies with chewing mouthparts, elongated antennae, and two pairs of elongated wings. Despite their large wings, they are poor and quite clumsy fliers. They undergo incomplete metamorphosis. Immature specimens are aquatic. Adults are often attracted to artificial lights at night.

Trichoptera (Caddisflies)
Resembling small moths, caddisflies are primarily nocturnal, weak-flying insects. They have two pairs of transparent wings that are covered in dense hairs. They have elongated bodies; long, thin antennae; and reduced or vestigial mouthparts. The larvae are aquatic and may be predators, herbivores, or scavengers. They undergo complete development. Adults are often attracted to artificial lights at night.

Plecoptera (Stoneflies)
Stoneflies are a primitive group of drably colored, soft-bodied insects. Adults have elongated and somewhat flattened bodies, long legs and antennae, two pairs of membranous wings, and two prominent cerci off the tip of the abdomen. They tend to be short-lived and

weak fliers. The nymphs are fully aquatic, requiring clean freshwater habitats. They undergo incomplete metamorphosis. Adults are often attracted to artificial lights at night.

Where to Look for Insects

Insects and their relatives can be found in virtually any terrestrial and freshwater environment and are generally common in and around the locations where people live. In other words, there are a great many insects and their relatives around to enjoy. However, due to their small size and often secretive habits, many insects can go unnoticed; that is, unless you know how and where to look. Observing and studying insects is like opening a treasure chest of natural history. You will quickly discover a hidden world filled with an amazing variety of form and function, including many unique interactions and bizarre behaviors. What is most exciting, though, is that this fascinating world is just outside your front door!

Of course, insects and their relatives occupy an amazing variety of ecological niches—a niche is the individual role and position an organism occupies in an ecosystem. Even a small space in the landscape or single object, such as a blooming plant or downed tree, can harbor a remarkable variety of species and offer hours of exciting exploration. In fact, many of the best places to look for this array of hidden biodiversity are under commonly encountered objects. The following provides a brief overview of where and how to explore these secret and often overlooked sites.

On Logs

Logs provide a bounty of resources for many insects and other arthropods. In many ways, they are a rich, miniature ecosystem alive with critters of all kinds. Insects may live on, in, or under the slowly decaying wood, but there is typically little outward evidence of this diverse system unless the seemingly lifeless log is examined closely and ultimately turned over. The organisms found on or under logs generally fall into one of four basic categories:

decomposers, predators, nesters, and hiders. Decomposers are organisms that feed on decaying dead organic material and in the process convert it to hummus, which in turn improves the soil and ultimately provides beneficial nutrients back into the system for other plants and animals to use. The following are a few decomposers: earthworms, sow and pill bugs, snails, beetle grubs, etc.

The diverse array of decomposers found under and around a decaying log attracts many potential predators that seek to take advantage of the available prey. Some examples include ground beetles, earwigs, and so on.

The next group of organisms found in a decaying log are nesters. Like miniature developers and architects, they tunnel through the log as well as the soil beneath, often creating an apartment-complex-like assemblage of chambers for their developing colony. Both termites and ants are nesters that inhabit dead wood. Their activity helps break down the log over time and provides additional holes, cavities and food for other organisms.

Logs also provide valuable shelter to many organisms. Many insects and arthropods are active at night and seek refuge by day in cool, dark, and often moist places. Such locations help minimize desiccation (drying out), provide protection from temperature extremes, and offer ideal hiding places to avoid detection by larger predators. Beyond such temporary occupation, logs also provide longer term habitat for hibernating organisms that seek to overwinter. Many wasps, bees, bugs, millipedes, and even some moth caterpillars, such as woolly bears, survive the cold winter months by hiding out in such sheltered locations.

How to Hunt for Bugs in Logs
So now that you know some of the organisms found on, in, and under logs, it's time to start exploring. It may be useful to have a magnifying glass and a wide-mouth plastic jar with a lid before heading outside. Once you find a log, take a few minutes to look it over carefully. Some organisms worth observing might be on or around the log. Then, carefully roll the log

over. Be prepared to look quickly, as many organisms will rapidly scurry when disturbed. Use the plastic jar to temporarily capture any critters for closer inspection. Studiously examine the underside of the log and the ground beneath for insects and other arthropods. You may wish to use the magnifying class to study some of the more minute organisms. When you are done looking, always gently return the log to its original position, so you don't significantly disrupt the valuable habitat. Then move on to the next log and continue the process. You will be amazed at all the critters you'll find!

In Leaf Litter

As you walk through a forest, take note of all the fallen leaves on the ground. This blanket of old, dead leaves and the associated twigs, discarded blossoms, fruit, and shed bark is actually a habitat all its own and home to a surprising variety of invertebrates. Such leaf litter is critical for creating healthy, moist, and nutrient-rich soil for the entire ecosystem. It also harbors a rich array of fungal and microbial life. Thus, leaf litter, just like a decaying log, serves as great habitat for a wide range of insects and other arthropods. Just like with logs, it provides food and protected, often highly moist sites that support decomposers, predators, nesters, and hiders, creating a rich microcosm of life. Earthworms, pill and sow bugs, some fly larvae, snails, slugs, earwigs, and many others feed on dead or decaying plant material and can be prevalent in and under leaf litter. Predators or scavengers found in leaf litter include wolf spiders, some adult beetles and their larvae, ants, millipedes, cockroaches, centipedes, and earwigs; many are active by night and spend the daylight hours hidden under leaves in protected and moist locations. Many of these same organisms nest or reproduce in this environment. Lastly, leaf litter literally provides layers of protection to temporarily shelter critters, enabling them to mitigate temperature fluctuations, avoid desiccation, and escape hungry predators. It also provides longer-term protected sites in which to successfully overwinter.

How to Look for Insects in Leaf Litter

Similar to a dead log, leaf litter initially appears lifeless until you look more closely. Once again, it may be useful to have a magnifying glass and a wide-mouth plastic jar with a lid, as well as a small garden trowel and a plastic bowl before you start. Find a spot in your yard or nearby woods and get down and simply watch for a bit. You may see or hear a variety of critters crawling over or in the leaf litter. Then, use your hands or the small garden trowel and gently start pulling away the layers of leaves and debris, watching closely at what you uncover. The plastic jar can be used to temporarily capture any small organisms for closer inspection. Continue to uncover the debris until you get to the decomposed organic material and soil beneath. Using the trowel, now slowly dig down taking a few small scoops and placing them in the bowl. This will allow you to inspect that moist material in greater detail to see what life lies inside. When finished, return the organic material and soil to the same location and be sure to release any captured critters. You may wish to repeat the process at night and simply observe the ground to see a wider array of active organisms.

Beneath Tree Bark

Dead trees, stumps, and decaying logs typically are covered with loosening bark. While many insects and other arthropods live in or under dead or decaying wood, some can also be found under bark. In many ways, bark provides the same sheltered environment and resources as leaf litter and the area beneath logs. Similarly, many of the same organisms can be found here. A quick word of caution, however: Be very careful around dead trees or large stumps, as they are often unstable. Limbs or large pieces can fall and cause serious injury.

How to Find Insects Under Tree Bark

Once a safe tree, stump or log has been located, start by observing the exterior for signs of life. Do you see any small holes or other signs of activity? Then, slowly and carefully remove pieces of loose bark to see what lies beneath. Look quickly, as many organisms will quickly scuttle away from the disturbance. A wide-mouth jar is again handy, as it enables you to examine your finds. Once the bark is removed, explore both the uncovered dead wood and the back of the bark as you may find organisms on both. Some common organisms that you may encounter include ants and their brood, centipedes, sow and pill bugs, small beetles, beetle grubs, and earwigs, among many others. If you explore these sites in fall and winter, you may also discover many hibernating insects, including lady beetles and bugs.

Beyond any living critters, you will likely also see a history of activity, this can include everything from small holes and chambers to a network of small tunnels and routes that were created by from bark beetles or other wood-boring insects. The network of designs can be quite extensive and appear almost like a piece of artwork. Use the magnifying glass to examine the surface in greater detail. Dead trees and larger stumps will continue decay to further and eventually end up on the ground; continue to follow and explore these hidden resources through time as they decompose and as new organisms use or colonize them.

Looking for Insects on or Under Other Objects

While logs, large fallen branches, and leaf litter are great resources for exploring the hidden world of biodiversity, many other natural or artificial objects provide shelter and habitat for insects too. These include rocks, old boards, landscape pavers, and even flower pots. As you explore, gently turn over many of the larger objects that you find. These are great places to spot earthworms, earwigs, sow and pill bugs, millipedes, centipedes, snails, and cockroaches. Once you're finished looking, gently return the object to its original position before moving on.

On the Ground

A great many insects and arthropods are ground-dwelling, and they spend all or some of their time foraging, nesting, or otherwise scurrying along the ground in either natural or artificial areas. The specific species and the overall diversity of organisms depend greatly on the type of environment and the level of disturbance present. For example, the organisms found directly around your home and yard are often much different than those found in nearby woods or meadows. Nonetheless, in pretty much any area, there are a number of interesting ground-dwelling critters to discover. A number of organisms, such as wolf spiders, tiger beetles, and ground beetles, are active hunters and move about in search of available prey. Other predators, such as ant lion larvae, build elaborate pit traps to capture passing organisms. Many others are considered scavengers, feeding on a variety of living and dead animal or plant material. These include various ants, earwigs, centipedes, millipedes, snails, slugs, pill and sow bugs, and others. Many are most active at night and spend their time foraging and looking around for available resources.

Ground-Nesting Insects

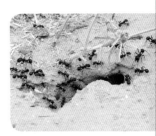

A great number or insects and other arthropods also nest in the soil. This includes some 70 percent of the native solitary bees in North America as well as many wasps and beetles. In turn, velvet ants search out and parasitize the nests of other insects, often ants and wasps. Fire ants and pavement ants are ground nesters, creating extensive subterranean colonies and actively foraging for food in the surrounding area. Carpenter ants typically nest in dead wood but commonly also excavate the soil beneath fallen trees or decaying logs. Yellow jackets take advantage of existing animal burrows or other cavities to form extensive underground nests. Mud dauber wasps frequent moist ground to acquire the raw materials to build their adobe-like nests. A number of organisms, such as earthworms and slugs, live much of their lives in the soil itself but can often be found aboveground periodically. Rainy days are a great time to spot earthworms, for example.

Some insects also feed at or near ground level. Male butterflies are one example; they regularly visit wet sand, gravel, mud puddles, or animal dung to gain nutrients. Many butterfly and moth larvae will periodically be seen wandering along the ground. When you see them doing that, they've finished feeding and are looking for suitable and protected places to pupate or spin a cocoon—or in search of places to overwinter, in the case of the woolly bear. In some cases, the larvae of various insects, including those of regal moths, imperial moths, and tomato hornworms, actually pupate in the soil.

While it may not be the first place you think of to look for insects and other arthropods, the ground is actually a habitat bustling with life of all kinds. Spend some time looking down and you will be amazed at what you find.

On Plants

Plants provide vertical structures in the environment, and they range from ground-hugging shrubs to trees that can reach hundreds of feet in the air. This, combined with the sheer diversity of plant species, provides a wealth of resources for insects—everything from food to shelter—and so plants support an almost countless variety of invertebrate organisms. A large percentage of insects and other arthropods are herbivores. They feed on, or in, every conceivable plant part, from stems, sap, and leaves to roots, flower buds, and fruit. We may be most familiar, though, with the critters that feed on leaves. Their activity is often extremely visible, resulting in noticeable leaf damage. However, it is important to note that the food preference of adult and immature insects often varies tremendously. Butterflies and moths are good examples. As caterpillars, they readily devour leaves and other plant parts; as adults, they feed exclusively on liquid food resources, such as flower nectar, tree sap or the juices from fermenting fruit. A variety of beetles, including tortoise beetles, Japanese beetles, may beetles, and grapevine beetles, also feed on leaves, while some of their larvae or grubs feed underground on plant roots. Grasshoppers, katydids, slugs, snails, and walking sticks are generalist herbivores, nibbling away on the leaves of various trees, shrubs, grasses, or herbaceous plants. Many bugs have piercing-sucking mouthparts that are used to pierce plant tissues and siphon out sap. These include treehoppers, tarnished plant bugs, green stink bugs, large milkweed bugs, leafhoppers, planthoppers, aphids, and cicadas, among many others.

Plant-Based Predators

The tremendous bounty of herbivorous critters attracts an equally rich diversity of predators seeking a hearty meal. These include active hunters, such as paper wasps, cicada killers, robber flies, jumping spiders, mud daubers, lady beetles, and spined soldier bugs, to various scavengers, including carpenter ants and fire ants. Plants also harbor several sit-and-wait ambush predators. The praying mantis is one example; it is one of the most charismatic carnivorous insects on the planet. Similarly, many spiders use plants as support structures for the large webs that help them capture insect prey. Still other insects may nest in or on plants, occupying shelter sites under leaves, in hollow stems, or in existing cavities. A few examples include paper wasps, leafcutting bees, and bald-faced hornets. They are equally useful as secure, protected sites on which to spin a cocoon or form a chrysalis. Plants can also provide the raw materials for nest construction, camouflage, or even body armor. Examples here include leaf cutting bees, which line their nests with pieces of leaves, or bagworms, which cover their bodies with sticks and other plant debris that they spin together with silk, creating a protective disguise.

How to Look for Insects on Plants

Carefully inspecting plants will reveal all sorts of unusual critters. Look for signs of feeding to start, and be sure to inspect all areas of the plant, not just the leaves. You can also put down a white sheet or a cloth beneath overhanging branches or taller plants and beat the vegetation with a stick. This will knock off many organisms onto the sheet below where they can be more closely observed or temporarily captured. You can also use an insect net to vigorously sweep through grasses and other vegetation and then look inside to see what critters have been dislodged.

Insects on Flowers

Flowers provide an abundance of sugary nectar and protein-rich pollen. These abundant and attractive food rewards attract insects, which help pollinate plants. In many ways, flowers are their own miniature ecosystems; they draw in a wide variety of insects from the surrounding environment, often in large numbers.

Predators Amid the Blooms

The resulting abundance of insects subsequently attracts other organisms with completely different intentions—to prey on the unsuspecting visitors to flowers. These include a wide variety of spiders, robber flies, assassin bugs, ambush bugs, stink bugs, and green lacewings, as well as larger, more charismatic predators, including praying mantises and even dragonflies. As many of these predators sit and wait for passing prey, they can be camouflaged, making them a challenge to spot. Careful and close observation is often needed.

Herbivores on Flowers

A variety of plant-eating insects can be found on or at least near flowers. They feed on the leaves, stems, buds and even flowers of their host plants. Within this mix, you may encounter various caterpillars, beetles, true bugs, aphids, thrips, and grasshoppers. Additionally, you may notice scavenging ants or the larvae of green lacewings and lady beetles, which are attracted to the aphids or scale insects that can be common on many plants.

Looking for Insects on Flowers

Without a doubt, flowers often teem with life and are a great place to start exploring. Before you look up close, stand back and observe for a few minutes. If the flowers are attractive, you should notice a fair amount of activity in the form of insects coming and going. Particularly noticeable insects include bees, wasps, butterflies, and day-flying moths. As you look more closely, you may notice a variety of smaller and less obvious critters, such as beetles and flies. The vast majority of these insects are attracted to the copious quantities of nectar and pollen, and they are generally classified as pollinators.

As you start looking, you will quickly notice that different flowers attract different organisms and that some species, types, and colors of flowers are much more attractive than others. Planting a flower garden in your yard or at your school is a great way to help provide food and habitat for many of these beneficial organisms, and it's an easy way to bring bugs to you, making observation easy.

At Lights

Artificial lights are a magnet for nocturnal insects. They are drawn in from the darkness and often linger until dawn near the light. Scientists aren't exactly sure why bugs are attracted to artificial lights. It may be that the light interferes with their natural navigational cues. It may also be due to the wavelengths that a particular light emits, as we know insects are more attracted to ultraviolet (UV) and short wavelength colors. Regardless, artificial lights are a sure-fire way to draw in bugs and an excellent place to start exploring.

What You'll Find

While lights attract a wide array of critters, what you'll find—and how many will show up—depends on many factors, including the time of year, the ambient temperature, the type of light, and your location. In general, insects are more active on warmer nights. Lights that produce ultraviolet light (often known as black lights) are more attractive and are widely used by entomologists. What you'll find also changes by season; nearby habitat and competing light sources matter too. As a rule, if your light's the only one, you'll attract more insects than if there is competition.

Attracting Insects With a Light

On a good night, a productive light can attract a huge number and variety of organisms. Even from a distance you can spot larger insects circling about or perched on an adjacent structure or wall. Moths are often the most numerous and obvious. They range greatly in size, from giant silkworm moths, such as the Luna Moth or Cecropia Moth, to very tiny micromoths, which deserve a much closer look. A great many beetles are also regularly attracted to light, including the common early season May or June beetles. Katydids, giant water bugs, roaches, green lacewings, dobsonflies, and antlions may be regular visitors too. If you are close to water, mayflies can be tremendously abundant at times, along with stoneflies and caddisflies. Beyond insects, lights often also attract some predators. Spiders often build webs near light sources to take full advantage of the abundant and easily accessible prey.

Observing the organisms that are attracted to artificial lights around your home is a good way to quickly learn about the wide variety of bugs near you. It will give you a quick snapshot of the usual suspects you can expect to regularly encounter. You can also set up black lights to more purposefully attract these nocturnal critters. In either case, regular observation at artificial lights is a fun and easy way to get to know your local insect community.

In or Near Water

Many insects live in or around bodies of water for some portion of their life cycle. Streams, ponds, rivers, and even lakes can provide ideal habitat and access to a wealth of food resources. However, surviving in aquatic environments requires many special adaptations to enable insects to breath, move, avoid predators, and find food. These differences make exploring the insect world around or in water all the more exciting.

Looking for Insects in Water

As you approach a stream or pond, move slowly and carefully watch for movement around the edges of the water and nearby vegetation. Dragonflies and damselflies are common and conspicuous inhabitants of this realm. They fly around the habitat, often scurrying low over the water to lay eggs, hunt for small flying insects, or perch on adjacent or emergent vegetation. Pause briefly near the water and watch the circus of activity. It can be quite amazing! Next, scan the surface of the water for motion. You may encounter clusters of whirligig beetles racing quickly in chaotic circles or water striders darting abruptly forward, as if they were literally walking on the water. Then look below the water surface. Here, you may see backswimmers submerged upside down, wriggling mosquito larvae, or diving beetles moving up and

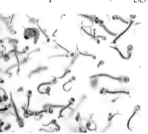

down in the water column. In shallow water or around pond margins you may also see dragonfly or damselfly niads on submerged vegetation or on the silty bottom. A small aquatic or aquarium net can be used to explore this word in greater detail. In many ways, this aquatic environment is truly a hidden world of excitement and discovery.

Beware of Bites and Stings

Many of the organisms described in this book can bite, sting, or pinch. They can often cause temporary pain, redness, itching, or minor swelling, and these injuries usually heal on their own. With that said, if you're allergic to a specific kind of insect bite or sting—bees, wasps, and ants are among the most common culprits—a bite or a sting can lead to severe reactions, which can have life-threatening symptoms and require emergency treatment. If you know you're allergic, or someone in your family is, don't closely approach, molest, or handle insects, and always be aware of your surroundings. Use caution when turning over logs or other objects, or when reaching into cavities or crevices, and wear gloves if you plan on doing so. Of course, children should always be under adult supervision.

Lastly, be very careful when walking near bodies of water. Rocks, branches, logs, or muddy banks can be slippery and dangerous. Adult supervision is always recommended. Remember: a safe experience is a fun experience.

True Bugs (Order Hemiptera)

At Lights

Giant Water Bug

Size: 1.75–2.25 inches long

ID Tips: Very large, dull brown wings overlap to form a distinctive X-like pattern on the back; forward-facing, pincher-like front legs

Range: Throughout the United States

With adults reaching nearly 4 inches in length, the giant water bug is the largest true bug in the United States and Canada. They are predatory, aquatic insects commonly found in ponds, slow-moving streams, and other freshwater wetlands with submerged vegetation. Hunting by ambush, a giant water bug sits motionless on plants just below the water surface and grabs passing aquatic organisms with its powerful front legs. It holds the prey firmly and pierces the victim with its sharp beak. It then injects enzymes that dissolve the body tissues and sucks up the resulting liquid meal. While they spend most of their lives in water, adults can fly and may move from one wetland to another in search of mates. During this time, they are frequently attracted to artificial lights at night. **Be careful,** though: Giant water bugs can give you a painful bite if handled.

Believe It or Not: Adults catch and eat insects, snails, tadpoles, small fish, and even frogs.

Alderflies, Dobsonflies, and Fishflies (Megaloptera)

At Lights

Eastern Dobsonfly

Size: 4.0–5.5 inches long
ID Tips: Very large, brown body with membrane-like gray-brown wings; males have huge jaws
Range: From the Rocky Mountains east

Almost prehistoric in appearance, the eastern dobsonfly is a ferocious-looking insect. The enormous adults approach 6 inches in length and have large, membranous wings. Nonetheless, they are relatively weak fliers and appear somewhat clumsy in the air. Males have long, curved mandibles that are about one-third the length of their body. Despite their monstrous form, they are completely harmless. By contract, females have relatively small but powerful mandibles and are capable of delivering a painful bite and even drawing blood. If disturbed, they assume a defensive posture by raising their head and flexing their mandibles, ready to fight off any attacker. Most active at night, the adults are readily attracted to artificial lights. Their predatory larvae are aquatic, living in fast-moving rocky streams where they prey on other insects. Good environmental indicators, eastern dobsonflies require clean water to thrive and are sensitive to pollution.

Believe It or Not: Dobsonfly larvae are commonly used as fishing bait and sold commercially.

Butterflies and Moths (Order Lepidoptera)

At Lights

Wavy-lined Emerald Moth

Size: Wingspan 0.5–0.65 inch
ID Tips: Wings green with fine scalloped, white lines
Range: Throughout the United States and southern Canada

This small moth is widespread and quite common across much of the United States and southern Canada. The delicate-looking adults are named for their uniform green coloration and fine white irregular lines. Like other geometer moths, the wavy-lined emerald holds its wings outstretched while at rest. They are commonly found at artificial lights where they are easy to photograph or closely inspect. The bizarre caterpillars are masters of disguise. Also called camouflaged loopers, they attach pieces of vegetation to their bodies; this helps them expertly blend in with the surrounding flowers on which they feed. Close inspection is needed to separate these tiny beasts from the blossoms.

Believe It or Not: The adorned larvae also occasionally sway back and forth, as if they were leaves gently blown by the wind. The resulting behavior helps to reinforce the nearly perfect deception.

Butterflies and Moths (Order Lepidoptera)

At Lights

The Beggar

Size: Wingspan 0.75–1.1 inches
ID Tips: Wings pale yellow with irregular dark spots; forewings rounded
Range: The eastern United States

This delicate moth has noticeably rounded wings and a somewhat weak, awkward flight. Overall, its appearance is quite unique and unlikely to be confused with any other moth. It is common in and near deciduous forests across eastern North America. Adults are regularly attracted to artificial lights at night. Unlike many other geometer moths, the beggar rests with its forewings held over the hind wings, forming the silhouette of an inverted heart. Few details are known about the moth's ecology or behavior. While the larvae are reported to feed on violets and possibly maples, it is likely that it has many other hosts.

Believe It or Not: While the exact origin is not known, the moth's odd name was possibly inspired by its irregular dark spots, which give the wings the appearance of tattered or hole-filled clothes.

Butterflies and Moths (Order Lepidoptera)

At Lights

Pale Beauty

Size: Wingspan 1.1–2.0 inches
ID Tips: Wings pale green with two dark-outlined pale stripes; hind wings with irregular borders and a stubby tail
Range: The northern two-thirds of the United States and all of Canada

Appropriately named, this small geometer moth has delicate wings that are pale green to almost whitish, and it holds its wings open while at rest. Common throughout much of the northern two-thirds of the United States and all of Canada, the adults are regularly attracted to artificial lights at night. It is a moth of forested habitats and adjacent open, shrubby habitats, although it can be regularly encountered in wooded subdivisions and urban parks. The elongated brown larvae feed on a wide array of trees and shrubs—from evergreens to broadleaved deciduous trees. Partially grown larvae overwinter. The pale beauty produces a single generation in northern latitudes and two farther south.

Believe It or Not: Unlike other geometer moth larvae, those of the pale beauty have unusual short hairs on the lower part of their bodies. This moplike fringe potentially helps aid in their defense, perhaps by breaking up the caterpillar's silhouette.

Rosy Maple Moth

Size: Wingspan 1.2–2.0 inches
ID Tips: Fuzzy yellow body with bright pink-and-yellow wings
Range: The eastern United States

The flamboyantly colored rosy maple moth is unmistakable. Generally common in deciduous forests across much of the eastern United States, the amount of pink on its wings is actually quite variable, with some individuals being more yellow. There is also an all-pale yellow or white form with little or no pink markings. The adults are strongly attracted to artificial lights and rest with their wings folded over their back in a triangular tentlike fashion. If touched, the adults typically fall to the ground, curl their abdomen, and play dead temporarily. Male moths tend to be smaller and have narrower, more pointed forewings compared to females. As their name suggests, the larvae primarily feed on maple trees and are gregarious when young.

Believe It or Not: Despite its gaudy coloration common to many unpalatable species, the rosy maple moth is apparently not toxic to predators. Some biologists have speculated that the moths mimic the showy winged maple tree seeds that are surprisingly similar in appearance.

Butterflies and Moths (Order Lepidoptera)

At Lights

Underwing Moth

Size: Variable; wingspan 1.5–3.0 inches

ID Tips: Variable; forewings typically dark with barklike pattern and hind wings with colorful bands

Range: Throughout the United States

This is a highly distinctive and diverse group of moths with more than 100 species found in North America. Adults have stout bodies and dark, dull-colored forewings with mottled or barklike patterns that they hold over their back while at rest. As a result, they are highly camouflaged when sitting on tree trunks, a common location to find them during daylight hours. However, when disturbed, underwing moths quickly spread their wings to reveal much brighter hind wings below before flying off to a nearby tree. Colored with bands of red, pink, yellow, orange, and even white, this hind wing pattern, coupled with their rapid wing motion, may help to startle interested predators. The showy nature of this group has made them popular with collectors and naturalists alike. Active at night, underwing moths are commonly attracted to artificial lights or may readily be drawn to sugar baits.

Believe It or Not: Underwing moths have simple ears that enable them to hear the ultrasound of a night-hunting bat. This early detection helps them avoid capture; to do so, they either move away from the approaching bat or fly erratically.

Butterflies and Moths (Order Lepidoptera)

At Lights

Giant Leopard Moth

Size: Wingspan 2.5–3.5 inches

ID Tips: Large; elongated white forewings with a mix of solid black and hollow black spots

Range: The eastern United States

With its bold white-and-black pattern, there is absolutely no mistaking this striking insect. The sizable giant leopard moths are common at artificial lights. If disturbed, they often drop to the ground and temporarily play dead. When doing so, they curl up their plump abdomens to reveal a bold-orange-and-iridescent-blue pattern, likely serving to scare off potential predators. If further molested, the moth secretes acrid yellow droplets from glands in its thorax. The large larvae are deep black with bright red rings on their bodies. They have a broad host range, feeding on a wide assortment of different plants and may readily move from one species to another. Fully grown larvae may reach three inches in length and are often spotted wandering along the ground in fall as they search for a protected site in which to overwinter.

Believe It or Not: Aptly named, the giant leopard moth is the largest tiger moth found in eastern North America.

Butterflies and Moths (Order Lepidoptera)

At Lights

Garden Tiger Moth

Size: Wingspan 1.75–2.7 inches

ID Tips: Forewings are brown with intricate white pattern; hind wings are orange with black spots; thorax brown; abdomen orange with black bands

Range: The northern half of the United States and into southern Canada

This is a wildly attractive moth of woodlands and adjacent open or shrubby areas, including gardens and yards. Due to its large size and elaborate color pattern, the garden tiger moth is a favorite among collectors and naturalists alike. It is found across southern Canada and the northern half of the United States. It also occurs throughout Eurasia. The distinctive hairy black larvae have a reddish brown coloring on the lower half and are commonly referred to as wooly bears, along with caterpillars of several other species. Its name alludes to the fact that the moth and its fuzzy caterpillars were common sights in European gardens. Active during the summer, the mature larvae are often seen wandering along the ground in fall where they seek protective sites in which to overwinter. The distinctive adult moths are attracted to artificial lights at night.

Believe It or Not: Unfortunately, this lovely moth has become a victim of climate change in many regions. Adapted to cold temperatures, the larvae have a challenge surviving mild, wet winters with little snowpack.

Butterflies and Moths (Order Lepidoptera)

At Lights

Io Moth

Size: Wingspan 2.0–3.0 inches
ID Tips: Mottled yellow to reddish-brown; a large eyespot on each hind wing
Range: From the Great Plains east

Colorful and distinctive, this medium-size moth is hard to mistake due to its large, target-shaped hind wing eyespots. When at rest, Io moths hold their heavily patterned and barklike forewings closed, concealing these prominent markings. If disturbed, however, they quickly fling their wings open to expose these realistic false eyes. This defensive maneuver may help to startle predators or at least deflect their attack away from the insect's vulnerable body. Io moth larvae are equally interesting. They are gregarious and feed together in small clusters throughout the majority of their development. Full-grown larvae are bright green with a bold red-and-black stripe on the side. The larvae are covered in short, branched venomous spines.
Be careful! If touched, they immediately generate a painful burning and itching sensation. While seldom more than just a nuisance, the pain and sensitivity in the affected area can last for several hours.

Believe It or Not: The Io moth is one of several caterpillars that have urticating "stinging" spines; most caterpillars that can sting produce moths.

Butterflies and Moths (Order Lepidoptera)

At Lights

Promethea Silkmoth

Size: Wingspan 2.8–3.7 inches

ID Tips: Large; wings black with light brown borders in males; wings two-toned in pink-brown with dark bases and a pale, wavy central stripe

Range: The eastern United States

The promethea silkmoth is arguably one of the most attractive large moths in eastern North America. Unlike most other giant silkworm moths, it is sexually dimorphic: males and females look radically different. Females are nocturnal and periodically come to artificial lights. However, mating does not occur at night. Instead, females begin releasing pheromones in the late afternoon. These volatile chemicals can travel long distances, and males are able to pick up the scent from many miles away. After a short amount of time, one or more blackish-colored males begin to fly in and locate the receptive female. Because of their large size and daytime activity, the large dark males are often mistaken for butterflies. Once mating is complete, females begin to lay their small whitish eggs in clusters on host leaves. The larvae feed on a broad range of trees, including wild cherry. The robust adult moths do not feed but instead rely on the fat reserves built up during the larvae stage.

Believe It or Not: Male promethean silkmoths mimic the toxic pipevine swallowtail for protection from predators.

Butterflies and Moths (Order Lepidoptera)

At Lights

Pandora Sphinx Moth

Size: Wingspan 3.0–4.5 inches
ID Tips: Elongated olive-green wings with pink streaks and darker patches
Range: The eastern United States

With velvety green wings and pinkish hues, the Pandora sphinx is a large and extremely spectacular moth. Generally common throughout much of the eastern United States, it is associated with mixed-deciduous forest but can be found in more suburban locations as well. The adults are most often encountered at dusk or dawn as they adeptly maneuver from blossom to blossom, almost like crepuscular hummingbirds. They have a long proboscis and can sip nectar from many long-tubed flowers, such as petunias or morning glories. The distinctive larvae are plump and chocolate brown with large, white circular spots along their bodies. They feed on the leaves of wild grape and Virginia creeper. When fully mature, they crawl down to the ground and pupate in the soil. One generation is produced in the north, and up to two occur in more southern locations.

Believe It or Not: Like many other moths, female Pandora sphinx release pheromones to attract males. They rest on vegetation while the interested males navigate to them.

Butterflies and Moths (Order Lepidoptera)

At Lights

Imperial Moth

Size: Wingspan 3.5–6.8 inches
ID Tips: Large; somewhat elongated yellow wings with a varying degree of purplish-brown markings
Range: The eastern United States

With females often having a wingspan of more than six inches, this impressive insect is one of the largest moths in North America. Its elongated yellow wings have varying amounts of purplish-brown markings and tend to resemble fallen leaves. This wing pattern likely provides effective camouflage, helping moths resting during the daytime blend unnoticed into the background vegetation. While adults are often attracted to artificial lights, males are more frequently encountered than females. The stout larvae may be either green or brown and are covered with fine hairs and have four short knobby horns behind the head. They feed on a wide range of trees, including oak, pine, maple, and hickory. When fully grown—and approaching some five inches in length—the larvae crawl down and wander extensively. Once a suitable location is found, they burrow into the soil and pupate underground.

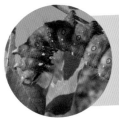

Believe It or Not: The sizable caterpillars produce extensive amounts of large, barrel-shaped fecal pellets called frass. These conspicuous droppings can often be spotted on the ground beneath an occupied tree.

Butterflies and Moths (Order Lepidoptera)

At Lights

Luna Moth

Size: Wingspan 4.0–4.5 inches
ID Tips: Large moth, light green wings with long hind wing tails and a furry white body
Range: The eastern United States

Truly a showstopper, the beautiful pale green luna moth is named for its round moonlike eyespots. Common in forested areas across eastern North America, the nocturnal adults are frequently encountered at artificial lights. The characteristic long, curved hind wing tails are more than just ornamentation. They actually help protect the moths from being eaten. As the moths fly, their tails flutter, producing an acoustic signal that confuses bats, common predators of night-flying insects. The resulting signal causes the bats to target the long tails, leaving the Luna moth's body unharmed. The chubby, bright-green caterpillars feed on a variety of hardwood trees, including walnut, hickory, persimmon, and sweetgum. When mature, they spin papery silken cocoons among growing leaves, which both eventually fall to the ground in autumn.

Believe It or Not: Adult Luna moths do not feed. Instead, they live off the food reserves acquired as caterpillars. The same is true for other giant silk moths.

Butterflies and Moths (Order Lepidoptera)

At Lights

Polyphemus Moth

Size: Wingspan 4.0–5.8 inches
ID Tips: Large; tan to reddish-brown wings with prominent hind wing eyespots
Range: Throughout the United States

This impressive insect is one of the largest and most widely distributed giant silkworm moths in North America. It may also be one of the most distinctive. No other large brown moth has the conspicuous yellow-centered black hind wing eyespots. Common in deciduous forests, it has adapted well to more urban settings, including wooded suburban neighborhoods, parks, and even parking lots that support its host trees. The adults are frequently attracted to artificial lights. Females are generally larger than males and have plump, round bodies and narrow antennae. Males have broad, fernlike antennae and tapered abdomens. The chubby, bright-green caterpillars feed solitarily on a wide variety of broadleaf trees, including oak, elm, birch, and dogwood, and may reach nearly three inches long when fully grown. The pale oval cocoons are about the size of a chicken egg. They are attached to branches with silk and frequently hang downward where they can be quite easy to spot on a leafless tree in winter.

Believe It or Not: Woodpeckers and even hungry squirrels often predate the robust brown pupae inside the egg-size cocoons.

Beetles (Order Coleoptera)

At Lights

May Beetle

Size: 0.5–1.0 inch long
ID Tips: Stout, unmarked shiny-tan-to-reddish-brown oblong body
Range: Throughout the United States

Also called June bugs, these abundant drab-colored beetles are virtually synonymous with the beginning of summer. Strongly attracted to artificial lights at night, they are clumsy fliers that often awkwardly bang into windows or screens with a droning buzz. Numerous species occur across North America, each with a distinct life cycle that varies in length from one to four years. Female beetles lay their eggs in the soil. The resulting larvae are C-shaped white grubs that feed underground on plant roots or decaying organic material. They are often encountered while digging in flowerbeds or under turf grass. A variety of common suburban animals—from skunks to opossums and moles—all feed on the white grubs, digging through loose or moist earth while they forage. When fully grown, the larvae pupate underground. As their name suggests, the rounded adult beetles begin to emerge the following spring in May and June.

Believe It or Not: In cold climates during the winter, the developing grubs will move deeper underground below the frost line to avoid freezing.

Beetles (Order Coleoptera)

Click Beetle

Size: 0.5–1.2 inches long
ID Tips: Brown to black; elongated often dull or unmarked body
Range: Throughout the United States

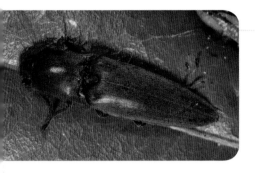

While often lackluster in appearance, click beetles have an entertaining personality. They are adept actors and readily play dead if disturbed. They accentuate this appearance by tucking their legs and antennae under their body and remaining motionless until danger has passed. Click beetles also have a much more exciting behavior. The beetle has a hinged joint on the thorax that allows it to arch back. Once in this position, it quickly snaps itself straight, causing a loud "click" and propelling the insect up in the air, often several inches high. This sudden leap, along with the accompanying sound, likely serves to effectively startle any would-be predator, including most humans. Adult beetles are nocturnal and are often seen at artificial lights. Their larvae mainly live in soil or rotting stumps where they prey on other insects or feed on plant material.

Believe It or Not: The distinctive clicking behavior can also help a beetle right itself if turned on its back.

Beetles (Order Coleoptera)

At Lights

Ten-lined June Bug

Size: 0.75–1.25 inches long
ID Tips: Brown oval body; vertical white stripes and noticeably clubbed antennae
Range: The western United States

This is a distinctive, stout beetle of the western United States. The large adults are brown with bold white stripes and are commonly attracted to artificial lights at night. They typically go unnoticed during the day, hiding in leaf litter or weedy vegetation. Male beetles have prominent fanlike antennae that they use to detect pheromones produced by females. Mated females lay eggs in the soil. The resulting whitish grubs feed on plant roots and typically take two years to fully develop. They can cause significant damage to various crops, including fruit trees and many other economically important plants. Depending on the geographic location and length of the growing season, it can take several years for the grubs to complete development.

Believe It or Not: The adults make a relatively loud hissing sound when handled or disturbed. This noise presumably startles attacking predators.

Beetles (Order Coleoptera)

Grapevine Beetle

Size: 0.8–1.2 inches long
ID Tips: Oval yellow or orange-brown body with black spots and legs
Range: The eastern United States

The grapevine beetle is a member of a diverse group of conspicuous, often quite showy, beetles called shining leaf chafers. Altogether there are about 4,000 species found worldwide. Throughout the eastern United States, the grapevine beetle, or spotted June bug as it is sometimes called, is commonly encountered at artificial lights. The bulky adults have a strong but somewhat clumsy flight and make a noticeable buzzing sound. As their name suggests, they feed on the leaves of grapes and Virginia creeper, two common and widespread plants in and near wooded areas. Their whitish larvae live in the soil and feed on rotting wood.

Believe It or Not: Because of their spotted appearance, they are sometimes mistaken for gigantic ladybugs.

True Flies (Diptera)

At Lights

Crane Fly

Size: 0.5–2.5 inches long
ID Tips: Slender body with extremely long, thin legs; two wings
Range: Throughout the United States

At first glance, crane flies may look like monstrous, titanic mosquitoes. Nevertheless, they do not bite and are completely harmless. In fact, most adults eat very little, if at all. These distinctive insects are named for their extremely long, delicate legs, which suspend their equally narrow bodies high above them, much like a crane. Adult crane flies have a slow, somewhat bouncing flight, especially when in your house or against an outside wall. They are readily attracted to artificial lights and may often be found in good numbers. Their larvae occur in a variety of moist environments. Some are fully aquatic, living in ponds or other wetlands, while others dwell in damp soil, moss, or dead wood, where they feed primarily on decaying organic matter.

Believe It or Not: Crane flies are one of the largest groups of flies in the world, containing more than 15,000 different species and subspecies.

Antlions, Lacewings, and Mantidflies (Order Neuroptera)

At Lights

Green Lacewing

Size: 0.4–0.7 inch long
ID Tips: Light green slender body; long antennae and four clear translucent wings with green veins
Range: Throughout the United States

Green lacewings have two very distinct lifestyles. Adults are delicate insects that mainly feed on pollen, nectar, or aphid honeydew. They are weak fliers and are often seen at artificial lights. Their larvae, on the other hand, are very similar in appearance to lady beetle larvae and are ferocious predators. They tenaciously consume a wide variety of soft-bodied insect prey, including everything from aphids and mites to leafhoppers and small caterpillars. They can even successfully attack insects larger than themselves. As a result, they are considered highly beneficial and are sold commercially for biocontrol use in gardens and greenhouses. Green lacewing eggs are particularly distinctive. Laid singly or in clusters, the eggs are individually attached to long, thin filaments, making them resemble small, white balloons.

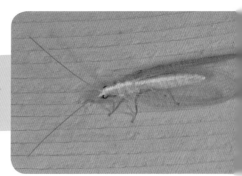

Believe It or Not: Some green lacewing larvae disguise themselves with debris, including plant material and food remains, to hide from predators.

Antlions, Lacewings, and Mantidflies (Order Neuroptera)

Ant Lion

Size: 0.8–1.5 inches long
ID Tips: Long, slender body with short antennae; large transparent wings
Range: Throughout the United States

Antlions get their name from their predatory larvae, which excavate deep pits to trap unsuspecting ants or other small insects. Resembling a cone-shaped depression, the pits are constructed in sandy soil and easy to spot. They often occur in groups when encountered. Most passing ants that fall in the pit rapidly slide down the unstable sides and into the waiting jaws of the larvae at the bottom. However, if prey tries to escape, the antlion throws sand particles at the ant to cause it to lose its footing and cascade down. Once captured, the antlion paralyzes its prey and injects enzymes to dissolve the tissues. It then drinks up the liquid meal and flings the old carcass out of the pit when finished. Adult antlions are also predators and consume a variety of soft-bodied insects. Primarily nocturnal, they are weak fliers but readily attracted to artificial lights.

Believe It or Not: Antlions are master architects and take many factors into account (such as diameter and wall slope) when designing their pits. This helps them optimize prey capture. They can even adjust their designs based on the overall texture of the sand.

Spiders (Order Araneae)

Goldenrod Spider

Size: 0.12–0.4 inches

ID Tips: Body yellow or white with or without a broad red stripe on each side of the abdomen

Range: Throughout the United States and into southern Canada

This abundant spider occurs throughout the United States and southern Canada. It is arguably the most frequently encountered crab spider. Individuals are ambush predators and do not spin webs. They sit and wait on blossoms for flower-visiting insects. Once in range, they grab the prey with their long front legs and inject venom from their fangs to paralyze the unfortunate passerby. They then suck out the resulting fluid. They are not limited to small prey and are fully capable of capturing larger insects, including bumblebees, grasshoppers, and butterflies. Its unique name comes from the fact that adults are often found on blooming goldenrod during the late summer and fall. Such plants are heavily visited by pollinators and provide a bounty of prey. Like many spiders, males tend to be significantly smaller than females, often less than one quarter the size.

Believe It or Not: The goldenrod spider can change color from white to yellow in order to effectively blend in with its surroundings.

Mayflies (Order Ephemeroptera)

Mayfly

Size: 0.25–1.0 inch long, not including tail filaments
ID Tips: Narrow pale to dark body; long legs; four wings held upright over the back; two to three long filaments off the abdomen
Range: Throughout the United States

Mayflies are delicate-looking primitive insects that exhibit two very different lifestyles. Their nymphs are completely aquatic, living at the bottom of fast-flowing streams or other freshwater habitats, where they feed on algae or other organic material. The nymphs often grow slowly and can take several years to reach maturity. When fully grown, they emerge from the water and molt into a preliminary winged stage before molting a second time shortly afterwards to become a fully reproductive adult. When they emerge, you'll know it, because they all do so at once, and the huge mating swarms are so large that they form huge dense clouds, can shut down roadways, and are sometimes even spotted on weather radar. It is during this time that large numbers are often attracted to artificial lights at night. Although called mayflies, adults can be found throughout the summer months.

Believe It or Not: Mayflies hold the record for the shortest adult reproductive life of any insect; most live less than one day.

Caddisflies (Order Trichoptera)

At Lights

Caddisfly

Size: 0.25–1.25 inches long
ID Tips: Narrow body with transparent wings; very long, thin antennae
Range: Throughout the United States

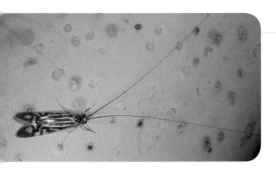

Caddisflies are small winged insects that resemble dull-colored moths. Although closely related to butterflies and moths, their wings are covered with fine, dense hairs instead of scales (as is the case with butterflies and moths). The adults are active fliers and are often found at artificial lights. When at rest, they hold their wings over their backs in a characteristic rooflike posture. Most tend to live less than two weeks. Caddisfly larvae are aquatic, inhabiting the bottoms of a variety of freshwater systems, ranging from rivers and fast-moving streams to ponds and lakes. Larvae may take several years to develop. Once they do, larvae often construct protective, tube-shaped cases by weaving twigs, sand, small stones, and other debris together with silk. The larvae carry this case along as they move and continue to expand it as they grow. After the larvae become adults, they leave the caddisfly case.

Believe It or Not: Some artists have taken advantage of the natural craftsmanship of caddisfly larvae by providing them with colorful stones or precious metals, which the larvae dutifully incorporate, creating an insect-made piece of jewelry. The resulting larval cases—manmade additions or no—are truly a work of art.

Stoneflies (Order Plecoptera)

At Lights

Stonefly

Size: Variable, 0.5–1.6 inches long
ID Tips: Dull brown; elongated and flattened body with transparent wings; long legs and two long tail filaments
Range: Throughout the United States

Adult stoneflies are dull-looking winged insects that are most often encountered near freshwater habitats. While they are typically poor fliers, good numbers can be attracted to artificial lights. During the daytime, look for them on low overhanging vegetation near rivers, streams, or lakes. Immature stoneflies, called nymphs, are aquatic and feed among rocks and gravel, subsisting primarily on submerged vegetation and algae. They prefer cool, fast-moving water that is well oxygenated and are intolerant of pollution. As a result, they are considered good environmental indicators of water quality. While adult stoneflies are short-lived, their nymphs may take several years to complete development.

Believe It or Not: Stoneflies, like many other aquatic insects, are fed upon by trout and other game fish. As a result, many of the lures used for fly-fishing closely resemble these insects.

Grasshoppers, Crickets, Katydids, Locusts, and Others (Order Orthoptera)

Mole Cricket

Size: 1.0–1.7 inches long
ID Tips: Brown elongated body with transparent wings; long hind legs; stubby front legs, two prominent tail filaments
Range: Mostly the eastern United States

By all accounts, mole crickets are not the most attractive insects. But what they lack in beauty, they make up for with function. They have streamlined bodies with highly modified and enlarged but stubby front legs perfect for digging. As their name suggests, mole crickets actively excavate tunnels and spend much of their lives underground. They are adept at moving in these restricted subterranean spaces. As a result, they can scurry quickly but are generally poor jumpers compared to most crickets. Mole crickets eat a variety of plant and animal material, feeding both aboveground and belowground on leaves, stems, or roots and even other insects. Several introduced species are considered serious pests. Adults can fly, typically to find mates or locate places to lay eggs, and are often attracted to artificial lights. Golf courses and other areas with turf grass are good places to find these unique insects.

Believe It or Not: Male mole crickets construct special burrows to attract females. The widened openings help amplify and project the mole cricket calls, which sound like a sort of cricket French horn.

Beetles (Order Coleoptera)

Whirligig Beetle

Size: 0.25–0.5 inch long

ID Tips: Shiny black, elliptical body; swims erratically on top of the water; often seen in groups

Range: Throughout the United States

Named for their rapid gyrations as they scoot across the surface of water like small speedboats, whirligig beetles are commonly found in the slowly flowing areas of ponds, lakes, and streams. They often occur in groups or "schools" that may contain dozens of individuals. The shiny black adults are a marvel of natural engineering. Active scavengers by day, they have divided eyes to see both above and below the water at the same time. They can quickly dive beneath the surface carrying a small bubble of air under their wing cases, like a scuba tank. They can also crawl on land and even fly from one wetland to another—this much mobility is an unusual feat for any aquatic insect. They may occasionally be attracted to lights. Their larvae are also predatory and typically hunt for food at the bottom of streams and ponds.

Believe It or Not: Whirligig beetles hold the record for the fastest measured speed of any swimming insect, yet they still retain the ability to make sharp turns.

Flies (Order Diptera)

In or Near Water

Mosquito Larva

Size: 0.1–0.2 inch long

ID Tips: Elongated brown segmented body; numerous hairs, but no legs; an enlarged thorax, and a long breathing siphon originating from the rear

Range: Throughout the United States

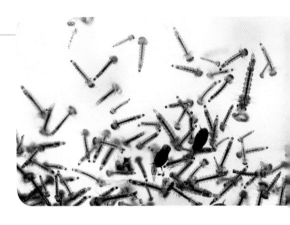

Mosquito larvae are fully aquatic. They are commonly found in many naturally occurring wetlands but prefer shallow water and thus can readily breed in everything from fountains and birdbaths to flower pots and old tires. Virtually any artificial container that can hold water is a possible habitat. Populations can be extremely dense, even in small vessels. The elongated larvae spend most of their lives close to the water surface. They breathe through a long tube called a siphon and dangle with their heads faced down, using their fanlike mouthparts to ingest algae, bacteria, and other microorganisms. Mosquito larvae develop quickly and are ready to pupate in just a matter of days. The winged adults emerge shortly thereafter. Mosquito larvae are important food resources for many other organisms, including aquatic insects, birds, and fish.

Believe It or Not: Mosquito larvae are often called wigglers due to their frequent back-and-forth thrashing movement in the water.

True Bugs (Order Hemiptera)

Water Boatman

Size: 0.2–0.45 inch long

ID Tips: Mottled brown; somewhat flattened oval body; large eyes; paddle-like hind legs

Range: Throughout the United States

These unusual aquatic insects are adept swimmers. They have strong, highly specialized hind legs that resemble oars, which help them propel themselves forward in a quick, somewhat jerky motion. By contrast, their front legs are short and adapted for collecting algae and other microorganisms. When not swimming, water boatmen typically anchor themselves to submerged vegetation. They are able to breathe underwater, much like a scuba diver; they do so by carrying a bubble of air around their abdomens. Water boatmen may be easily confused with the similar-looking and often equally common backswimmer, but they have flatter-looking bodies, tend to maneuver lower in the water column, and swim upright.

Believe It or Not: Water boatmen can be found in a wide range of aquatic habitats, including fresh, brackish, and salt water.

True Bugs (Order Hemiptera)

In or Near Water

Backswimmer

Size: 0.3–0.6 inch long

ID Tips: Light brown, somewhat flattened elongated body; large eyes and paddle-like hind legs; swims upside down

Range: Throughout the United States

This distinctive aquatic bug is a common sight in freshwater ponds, wetlands, or lakes. Adults can fly and are periodically attracted to artificial lights at night. As a result, they may even occasionally be spotted in swimming pools or birdbaths. As their name suggests, backswimmers maneuver upside-down using their long and powerful hind legs, which function like oars on a rowboat. Similar in appearance to water boatmen, they tend to be somewhat larger and are found at or near the water surface. Additionally, backswimmers are predatory insects, often resting near submerged vegetation to ambush passing prey. They use their short front legs to capture other insects, tadpoles, or even small fish and their piercing-sucking mouthparts to inject digestive enzymes and drink out the liquefied body contents. **Be cautious:** Backswimmers can inflict a painful bite if handled.

Believe It or Not: Backswimmers typically overwinter as adults. In northern regions during winter, they may even be spotted swimming under ice on a frozen pond.

True Bugs (Order Hemiptera)

In or Near Water

Water Strider

Size: 0.5 inch long

ID Tips: Brown narrow body with long middle and hind legs; found on the surface of water

Range: Throughout the United States

These amazing insects effortlessly move across the surface of water, much like miniature skaters. To do so, they have highly specialized legs that are both elongated to help distribute the insect's weight; their legs are also covered in fine hairs that repel water. The middle pair of legs helps propel the insect, while the rear legs aid in steering. The front pair of legs is smaller and used to grab food. Water striders are predators or scavengers that feed on living and dead insects, including many that accidentally fall into the water. They prefer calm areas in ponds, streams, marshes, or other wetlands and are often seen in groups. Because of their small size and delicate bodies, water striders are easy to overlook. Once spotted, though, they are very entertaining to watch but will scurry away quickly if approached too closely.

Believe It or Not: Water striders are extremely sensitive to vibrations on the water surface; they detect such vibrations to locate potential prey.

Dragonflies and Damselflies (Order Odonata)

Naiads (Dragonfly and Damselfly Larvae)

Size: 0.05–2.5 inches long
ID Tips: Flattened brown body with large eyes and a long abdomen
Range: Throughout the United States

Naiads are the immature larval stage of dragonflies and damselflies. Unlike the terrestrial winged adults, naiads are fully aquatic. After hatching from eggs, they live entirely underwater, molting up to 15 times as they grow. When fully mature, they crawl out of the water on a plant or stick to molt one last time into an adult. The old nymphal skins can often be found on emergent vegetation. These odd-looking creatures are voracious ambush predators. They wait at the bottom of wetlands or within submerged vegetation for passing prey. They have specialized mouthparts consisting of an extensible, hinged lower jaw with small hooks that can rapidly shoot out, often extending nearly the entire length of the body. They use these hooks to capture other aquatic invertebrates, tadpoles, and even small fish or amphibians. The slender naiads of damselflies have three elongated leaflike gills off their abdomen for oxygen exchange, while the more robust dragonfly larvae lack the noticeable projections and have internal rectal gills instead.

Believe It or Not: Dragonfly naiads can rapidly contract their rectal muscles, causing water to shoot out their rear ends. This unique jet propulsion rapidly propels the organisms forward underwater.

Dragonflies and Damselflies (Order Odonata)

In or Near Water

Eastern Amberwing

Size: 0.8–1.0 inch long
ID Tips: Small; brown body with transparent orange-to-amber wings
Range: The eastern United States

The eastern amberwing is a small dragonfly common throughout much of eastern North America. In fact, it is the second-smallest dragonfly in the United States. As its name suggests, the eastern amberwing has short, somewhat stubby wings that are noticeably orange to amber in color, particularly in males. Females have somewhat lighter-colored wings with large dark brown patches. Adults frequent small ponds, roadside ditches, or other wetlands with still water; they scurry quickly over the surface in a somewhat frenetic fashion. They regularly perch on branches or other emergent vegetation. While males often stay close to bodies of water and their established territories, females are known to wander quite far into neighboring areas and are often seen in adjacent meadows or fields. Mated females fly low across the water surface and tap their abdomens to lay eggs. The aquatic nymphs, called naiads, are predators of other small wetland invertebrates.

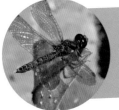

Believe It or Not: The eastern amberwing's banded abdomen mimics a wasp's appearance. This disguise is further enhanced when the amberwing moves its wings and abdomen up and down—like a wasp—when perched.

Dragonflies and Damselflies (Order Odonata)

In or Near Water

Halloween Pennant

Size: 1.0–1.6 inches long
ID Tips: Amber-orange wings with broad, dark-brown to black bands and spots
Range: From the Great Plains east

The Halloween pennant is one of the most distinctive and easily recognized dragonflies east of the Rocky Mountains. It is named for its bold light-orange-and-black wing pattern, which is reminiscent of the customary colors of Halloween. It is common to locally abundant in wetlands with emergent vegetation, on which the adults prefer to perch. They may also be seen in nearby fields and open weedy sites, where they forage for insect prey often some distance from water. Adult Halloween pennants have a somewhat fluttering, dancing flight that is similar to that of butterflies. They also perch differently than most other species, holding their forewings at a higher angle than that of their hind wings. This posture is quite noticeable even from afar.

Believe It or Not: There is some speculation that the Halloween pennant's color mimics that of a monarch butterfly and is intended to mislead predators. This potential deception could be reinforced by its more relaxed, fluttering flight behavior.

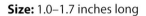

Dragonflies and Damselflies (Order Odonata)

Pond Damselfly

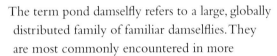

Size: 1.0–1.7 inches long

ID Tips: Variable color; thin elongated body with transparent wings held together over back; head with large eyes

Range: Throughout the United States

The term pond damselfly refers to a large, globally distributed family of familiar damselflies. They are most commonly encountered in more still-water environments, such as ponds and marshes, although some species also frequent moving streams. They are delicate creatures with conspicuously thin, elongated bodies that are often hard to easily follow when in flight. Their colors vary tremendously between species, sex, and occasionally age, but many are brightly marked and often have some elements of blue, brown, or red. As a result, accurate field identification can be a challenge. Most pond damselflies hold their narrow transparent wings together over their backs when at rest, often perching on wetland vegetation or dead twigs. Like dragonflies, adult damselflies are predatory and feed on other small insects. Their nymphs are aquatic and equally adept at catching prey.

Believe It or Not: Mating pairs remain together in flight while the female lays eggs in or near wetland vegetation.

Dragonflies and Damselflies (Order Odonata)

In or Near Water

Common Whitetail

Size: 1.6–1.9 inches long

ID Tips: A stout body with clear wings and a wide central black band; abdomen is chalky white

Range: Throughout the United States

The common whitetail is a familiar and widespread dragonfly found across much of North America. The male is quite distinctive in appearance with black-banded transparent wings and a noticeably wide and chalky white abdomen that is easily visible from some distance away. In contrast, females have three-spotted bands on their wings and brown abdomens with white dashes on the side. Like other dragonflies, common whitetails are active aerial predators of small flying insects, including mosquitoes. They typically maneuver in search of prey over ponds, streams, and other wetlands with slow-moving water, and they can sometimes be seen in good numbers. They also frequently perch on low vegetation, downed branches or logs, or even on bare ground near their wetland haunts.

Believe It or Not: Male common whitetail dragonflies readily establish and aggressively defend territories, primarily for access to potential mates. Males venturing into an established territory are rapidly pursued by a defending male, which holds its white abdomen raised as a signal to the intruder.

Dragonflies and Damselflies (Order Odonata)

In or Near Water

Eastern Pondhawk

Size: 1.5–2.0 inches long

ID Tips: Males with a powdery blue body and unmarked transparent wings; females with a green body and dark abdominal spots and unmarked transparent wings

Range: In the United States and Canada from the Rockies east

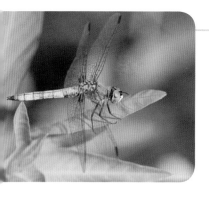

This is a common and widespread dragonfly found throughout the United States and Canada from the Rocky Mountains eastward. Often abundant around or near ponds, lakes, or other wetlands with calm waters, they regularly venture into nearby habitats and yards. Like other dragonflies, the sexes and immature adults often differ considerably from one another, almost looking like a completely different species. Mature males of the eastern pondhawk are an attractive powdery blue color with a green face. Females and immatures are bright green with broad, dark patches on the abdomen. Both sexes readily perch on vegetation or even on the ground and fly out to capture a wide assortment of small insects. When perched, adults are not overly wary and can often be easy to closely approach or observe.

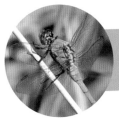

Believe It or Not: The common pondhawk has one of the longest flight seasons of any dragonfly, with adults often being active well into late fall.

Dragonflies and Damselflies (Order Odonata)

In or Near Water

Twelve-spotted Skimmer

Size: 2.0–2.25 inches long

ID Tips: Wings transparent with alternating dark and white spots; abdomen grayish

Range: Throughout the United States and southern Canada

This is a large, widespread, and often abundant dragonfly found across the United States and southern Canada. The insect's name refers to its prominent twelve dark spots; there are three on each of its four wings. Males also have white wing spots. The twelve-spotted skimmer prefers ponds, lakes, and other wetlands with emergent vegetation and more open, sunny, and exposed shorelines. Males are powerful and active fliers, regularly flitting back and forth along the water's edge and aggressively defending their territories from rivals or even other dragonfly species. While not a frequent percher, they do occasionally pause on prominent rocks, twigs, or other vegetation.

Believe It or Not: Although dragonflies have six well-developed legs, they are unable to walk. Their legs are well adapted to catch flying insects, however.

Dragonflies and Damselflies (Order Odonata)

In or Near Water

Common Green Darner

Size: 2.75–3.2 inches long

ID Tips: Large; thorax green with long slender blue abdomen; clear, transparent wings

Range: Throughout the United States

The common green darner is a large and often abundant dragonfly throughout North America. The sizable adults often have wingspans of more than 4 inches. Typically found near ponds and other slow-moving bodies of water, they are active, strong-flying predators that use their long legs much like a basket to capture a wide assortment of flying insects. While mosquitoes and other tiny insects form the basis of their diet, their size enables them to also feed on larger prey items, including butterflies, mayflies, bees, and even other dragonflies. There are even reports of common green darners capturing hummingbirds, although such reports are largely unconfirmed. Their alien-looking nymphs are fully aquatic. They use their specialized mouthparts, which consist of a lengthened and hinged lower jaw that extends outward, to grab various organisms, including other insects, tadpoles, and small fish. When fully extended, the lower jaw, or labium, may be nearly the length of the insect's body.

Believe It or Not: The common green darner is a well-known migratory dragonfly. Each fall, individuals from northern areas move southward, often in huge numbers, to overwinter in warmer climates.

In the Air

True Flies (Order Diptera)

Mosquito

Size: Typically less than 0.25 inch long

ID Tips: Narrow, often striped body with two wings; long thin legs and a prominent proboscis

Range: Throughout the United States

Well known for their high-pitched buzz and itchy bites, mosquitoes are among the most irritating backyard insects around. Actually small flies, they are a widespread and diverse group with more than 3,500 species worldwide. The adults are active fliers and feed on flower nectar and other sugary solutions. However, female mosquitoes require blood to produce eggs. Like miniature vampires, they seek out potential targets using a combination of vision, smell, and temperature. They then land and puncture the skin with their mouthparts. In the process, they inject saliva, which contains an anticoagulant to facilitate blood flow. The resulting itch is caused by your body's reaction to this foreign substance.

Believe It or Not: Mosquitoes breed in or near standing water and only require about a teaspoon of it to reproduce. The average backyard can produce thousands of mosquitoes each week, so be sure to regularly eliminate sources of standing water in your backyard.

True Flies (Order Diptera)

In the Air

Deer Fly

Size: 0.25–0.4 inch long

ID Tips: Variable color; brown to yellow or black body; large, often brightly colored eyes; banded wings

Range: Throughout the United States

These small, often yellow-and-black-striped flies are typically considered a nuisance. Often encountered in somewhat shady areas within or adjacent to wooded habitats, they actively attack passing mammals, including cattle, deer, dogs, and humans. They can be tenacious creatures and repeatedly return to bite the same victim over and over again if swatted away. Deer flies are attracted by body heat, carbon dioxide, and visual signals, such as motion. Like mosquitoes, female flies require a blood meal to produce viable eggs. They use their knifelike mouthparts to cut the victim's skin and then lap up the flowing blood. As a result, bites can be quite painful. By contrast, male deer flies do not bite but instead consume flower pollen and nectar. Females lay their eggs on vegetation near ponds, streams, or other wetland habitats. The resulting larvae are aquatic and typically feed on decaying organic material or small organisms in the wet soil and sediment.

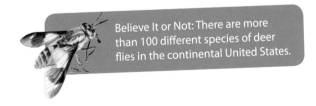

Believe It or Not: There are more than 100 different species of deer flies in the continental United States.

Beetles (Order Coleoptera)

Firefly

Size: 0.5–0.8 inch long
ID Tips: Flattened black body with orange and yellow markings; underside tip of abdomen is pale and produces light
Range: The eastern United States

Despite being called fireflies or lightning bugs, these distinctive insects are actually beetles. A common sight on summer evenings, fireflies literally light up the night by producing a chemical reaction inside a highly specialized organ in their bodies; the scientific name for this is bioluminescence. Each species of firefly produces its own distinctive flash pattern and some even flash simultaneously, creating a one-of-a-kind display. In general, males fly about, flashing as they go, while females flash back in response from perches on vegetation. The end goal of this elaborate display is to locate a suitable mate. However, some firefly species have learned to use this display for a more devious purpose. These females imitate the flash pattern of others to lure in unsuspecting males and eat them. Fireflies tend not to be active in highly illuminated areas, so look for them in dark, open areas away from artificial lights.

Believe It or Not: All firefly larvae also produce light and are commonly called glow worms. Fireflies are well defended chemically and known to be quite toxic if ingested by many predators. In fact, they can even be lethal to many reptiles and amphibians.

True Flies (Order Diptera)

On Flowers

Lovebug

Size: 0.25–0.35 inch long
ID Tips: Small dull black with orange thorax and long legs
Range: The southeastern United States

Lovebugs are a common sight during the summer months across the southeastern United States each year. Actually small flies, these bizarre insects are closely related to mosquitoes and may be just as annoying to many people. The adult flies are typically seen while mating, hence the name "love" bug, and the male and female lock onto one another and remain together for several days, even while flying. They appear twice a year in highly synchronized flights that can produce huge numbers, often resembling a biblical plague. During this time, they can be quite a nuisance, making their way into houses or splattering against the windshields of motorists. Fortunately, the adults do not bite or sting and live for only a few days, just long enough to find a mate and lay eggs. The maggot-like larvae feed on decaying plant material and are beneficial decomposers, helping recycle valuable nutrients back into the soil.

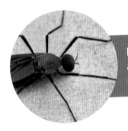

Believe It or Not: During large outbreaks, lovebugs are so numerous that they can actually clog vehicle radiators and can disrupt traffic.

True Flies (Order Diptera)

On Flowers

Flesh Fly

Size: 0.4–0.6 inch long

ID Tips: Gray body with longitudinal black stripes, a bristled abdomen, and reddish eyes

Range: Throughout the United States

These common flies are found throughout North America. They are similar in appearance to houseflies but typically a bit smaller. Flesh flies get their less-than-appealing name from the fact that their larvae or maggots typically feed on decaying material, dung, dead organisms, and open wounds on animals. Because of this lifestyle, they often quickly colonize human remains and thus are particularly useful in forensic investigation, helping to determine critical facts such as the time from death to corpse discovery. The larvae of some species are beneficial and feed on a variety of potential pest insects or other invertebrates. The drab bristly adults feed on nectar, sap, and other sugary substances using their sponge-like mouthparts to lap up the liquid meal. They may be frequently encountered at flowers.

Believe It or Not: Flesh flies do not lay eggs. Instead, the eggs hatch inside the female and she then deposits young maggots directly on available food.

True Flies (Order Diptera)

On Flowers

Bee Fly

Size: 0.4–0.75 inch long

ID Tips: Hairy body with wings held outward at rest; a long forward-pointing proboscis

Range: Throughout the United States

This is a diverse group of primarily small, brightly colored, and often fuzzy-looking flies that closely resemble bees. They are often seen hovering low to the ground in open, sunny areas but are capable of much more rapid and adept movements. Many bee flies have a long, stiff proboscis that protrudes out in front of their bodies. They use this feeding tube to sip sugary nectar from available flowers, much like a hummingbird. This behavior may help them avoid many sit-and-wait predators that reside on flowers and regularly ambush visitors when they land to feed. Despite their cuddly appearance, bee flies have a somewhat sinister lifestyle. Their larvae are carnivorous, feeding primarily on the eggs or other immature stages of other insects, from solitary bees and wasps to grasshoppers and tiger beetles.

Believe It or Not: Bee flies don't lay their eggs on a host. Instead, females deposit eggs singly on the ground, and the larva navigates to the nearby prey. To do so, it often crawls down into underground nest or burrows.

Beetles (Order Coleoptera)

Tumbling Flower Beetle

Size: 0.15–0.2 inch long
ID Tips: Small, rounded teardrop-shaped body with a pointed abdomen
Range: Throughout the United States

This is a diverse group of small, primarily dull-colored beetles that are frequently encountered on flowers. The adults have a distinctive humpbacked appearance with a noticeably tapered abdomen. They feed primarily on pollen. When disturbed, they use their powerful hind legs to leap from vegetation in a tumbling motion. This behavior is particularly noticeable when they are captured in an insect net. Tumbling flower beetles are active insects and readily fly from one location to another. Female beetles have a noticeably pointed ovipositor that extends off the abdomen; it is used to insert eggs into vegetation. The resulting larvae are primarily stem borers, but some species may also utilize fallen and decaying wood.

Believe It or Not: There are more than 200 species of tumbling flower beetles in the United States and Canada.

Beetles (Order Coleoptera)

Wedge-shaped Beetle

Size: 0.15–0.6 inch long

ID Tips: Compact, often stout, and somewhat rounded bodies; short wing coverings; often have fanlike antennae

Range: Throughout the United States

This is a highly unusual group of beetles, and relatively little is known about them. Much of what is known comes from a few species and involves a unique type of parasitism on bees and wasps. As the name implies, they are somewhat pie slice-shaped with a rounded back narrowing to the tip of the abdomen. Most noticeably however, are their short wing coverings called elytra that leave much of their wings exposed. Adults are often encountered on flowers. Following mating, females lay a small number of eggs on developing flowers. After hatching, the small larvae wait for a passing bee, grab hold of the insect, and hitch a ride back to the nest. The larvae feed internally on the developing bee brood and eventually consume the bee pupae.

Believe It or Not: Adult wedge-shaped beetles have very short life spans, with most surviving for only a day or so.

Beetles (Order Coleoptera)

Delta Flower Scarab

Size: 0.3–0.4 inch long
ID Tips: Wing cases are orange-brown with black markings; there is a prominent pale triangle on the thorax; hind legs are noticeably long
Range: Primarily the southeastern United States

The delta flower scarab is a small, colorful beetle that is named for the distinctive triangular-shaped mark on its thorax that also represents the letter delta in the Greek alphabet. Active during the daytime, the adults are regularly spotted on flowers where they feed on pollen and may be particularly abundant in late summer and early fall. They tend to be fond of dense flower heads with a proliferation of small individual blooms. It is also common to see mating pairs on blossoms. The delta flower scarab occurs throughout much of the eastern United States from Illinois south to Florida. Their larvae or grubs feed on decaying wood.

Believe It or Not: When disturbed, adult beetles lean forward and raise their long legs in the air. This unusual posture showcases its namesake Delta mark, which is thought to resemble the head of a stinging insect, such as a wasp, in order to deter potential predators.

Beetles (Order Coleoptera)

On Flowers

Blister Beetle

Size: 0.35–1.25 inches long

ID Tips: Variable in color; elongated cylindrical body with soft leathery wing covers somewhat rolled over the abdomen

Range: Throughout the United States

This is a diverse group of soft-bodied beetles with a distinctly narrow neck. Unlike many other beetles, their wing covers curve over their abdomens forming a distinct cylindrical body. Blister beetles are often conspicuously colored and easy to spot. Adults are plant feeders and are especially fond of flowers where they partake in both pollen and nectar. They can be particularly abundant in late summer or early fall when numerous individuals can often be found on a single blossom. Some adults also feed on vegetation. Their larvae are specialized predators, attacking other insects, including grasshopper eggs, immature bees, or the provisions in native solitary bee nests. Because of their reclusive lifestyle, blister beetle larvae are seldom encountered. If you encounter a blister beetle, be careful. They boast impressive chemical protection and can blast out a toxin that can cause blisters on the skin, especially if the bugs are inadvertently crushed.

Believe It or Not: Livestock feeding on plants containing large numbers of blister beetles can actually become severely ill and may even die from high levels of the toxin.

Beetles (Order Coleoptera)

Soldier Beetle

Size: 0.45–0.6 inch long

ID Tips: Highly variable in color and pattern; elongated soft wing cases that do not completely cover the tip of the abdomen

Range: Throughout the United States

Soldier beetles are narrow, streamlined insects often called leatherwings due to their relatively supple wing cases. They are similar in appearance to fireflies, which they are related to, and often mistaken for, but soldier beetles don't have the ability to produce light. Active by day, soldier beetles are readily encountered at flowers and can often be quite abundant, particularly in summer and early fall. They do not hesitate to fly, moving from plant to plant, albeit in a somewhat plodding manner, as is the case for many beetles. They can be effective pollinators. The adults of many species are predatory and feed on a wide variety of other small insects, while others prefer pollen and nectar. As a result of their diet and behavior, most soldier beetles are considered quite beneficial. Female soldier beetles lay their eggs in the ground. There, their carnivorous larvae aggressively hunt for small invertebrates in leaf litter or loose, moist soil.

Believe It or Not: Adult beetles defend themselves when threatened by secreting foul-tasting chemicals from specialized glands on their bodies. The distinctive coloration of many species helps alert potential predators that they possess this effective means of protection.

Beetles (Order Coleoptera)

On Flowers

Long-horned Beetle

Size: 0.5–2.5 inches long
ID Tips: Variable in color; slender oval bodies with noticeably long antennae
Range: Throughout the United States

Long-horned beetles are a charismatic group of often large and colorful beetles that get their name from their prominent, long antennae, which may exceed the body length of some species. They are incredibly diverse in color and pattern, ranging from uniformly dark to vibrantly marked. Some even mimic stinging insects, such as bees and wasps. Many boldly patterned species are active by day and are often found feeding at flowers, while others may readily be attracted to artificial lights at night. In either case, the adults tend to be strong fliers. Like metallic wood-boring beetles, female longhorns seek out weakened or damaged trees on which to lay eggs. Upon hatching, the larvae bore into the wood and complete their development inside the host tree. The larvae of both groups may be found inhabiting the same tree.

Believe it or not: Longhorn beetle larvae are very loud when they feed and may be actively heard chewing inside the infested wood.

True Bugs (Aphids, Cicadas, and Others)

Jagged Ambush Bug

Size: 0.4–0.5 inch long
ID Tips: variable mottled color pattern; violin-shaped body with enlarged grasping front legs
Range: Throughout the United States

Aptly named, the jagged ambush bug's distinctive violin-shaped body has irregular, ragged margins. A highly camouflaged sit-and-wait predator, it rests motionless on flowers, blending in perfectly with the background. When insects inevitably arrive to partake in the nearby pollen and nectar, the jagged ambush bug uses its powerful front legs to grasp the insect, much like a praying mantis. It then quickly immobilizes the prey by injecting digestive enzymes from its sharp, beaklike mouthparts and sucks out the liquefied body contents. Despite its relatively small size, it is able to capture insects much larger than itself, including honeybees, bumblebees, wasps, and butterflies. Adults are particularly abundant on goldenrod in late summer but may be found on a variety of different blossoms. Numerous individuals can often occupy the same flower.

Believe It or Not: Adult jagged ambush bugs are often encountered while mating, during which it is not uncommon for both the male and female to each have a captured insect.

Wasps, Bees, Ants, and Sawflies (Order Hymenoptera)

On Flowers

Leafcutting Bee

Size: 0.25–0.9 inch long
ID Tips: Variable; typically stout hairy black bodies with some lighter markings
Range: Throughout the United States

While these small solitary bees often go unnoticed, their presence in the landscape is unmistakable. Leafcutting bees, as their name implies, neatly cut semicircular penny- to quarter-size pieces out of leaf edges, making the leaves look as if they'd been put through a hole puncher. They then use this material to line their nests, which are constructed in dead wood, hollow twigs, or in the ground. Once the nest is constructed, the female bee will actively forage for pollen and nectar at available flowers. Unlike other bees, which carry pollen on their hind legs, leafcutting bees have specialized hairs on the underside of their abdomens for this purpose. As a result, adults often have noticeably yellow bellies. The female bee then brings the pollen back to the nest, mixes it with nectar and saliva into a small ball, places it into the nest and lays an egg. She seals off the chamber or cell and repeats the process until the entire nest is outfitted. Each resulting larva will feed on the pollen ball and complete development. Newly emerged adult bees overwinter, emerging in spring.

Believe It or Not: Like other solitary bees, leafcutters are not aggressive and rarely sting. Artificial nest boxes can be placed in a yard or garden to encourage breeding of these highly beneficial pollinators. See page 219 for one such project.

Wasps, Bees, Ants, and Sawflies (Order Hymenoptera)

Tiphiid Wasp

Size: 0.3–1.0 inch long
ID Tips: Body black with yellow bands, amber wings, and a long narrow abdomen
Range: Throughout the United States

This is a diverse group of primarily solitary wasps. The sexes are dimorphic; females are generally larger than males, which are long and slender. The overall color pattern of both is similar. Males are avid flower visitors and feed on nectar. By contrast, females tend to spend more time on or near the ground, searching for available prey. Tiphiid wasps are parasites of subterranean beetle larvae; they crawl into underground tunnels until locating a host. Once found, the female wasp partially paralyzes the grub with its sting and lays an egg on it. The resulting wasp larva then proceeds to consume the host, completing development in the process.

Believe It or Not: Tiphiid wasps are considered highly beneficial as pollinators and as natural enemies of invasive species. In fact, some species have been used as biological control agents to attack destructive Japanese beetles and other pest beetles.

83

Wasps, Bees, Ants, and Sawflies (Order Hymenoptera)

On Flowers

Metallic Green Sweat Bee

Size: 0.4–0.5 inch long

ID Tips: Head and thorax are metallic green with dark wings; abdomen metallic green or yellow with black bands

Range: Throughout the United States

When it comes to bees, these members of the genus Agapostemon are arguably some of the most attractive. Aptly named, the relatively diminutive adults are iridescent green with varying degrees of yellow and black banding. Metallic green sweat bees are avid flower visitors and are important pollinators. They nest in the ground, building elaborate tunnels for their developing brood. Females collect pollen, amass it together with nectar to form a large ball, place it in an individual cell, and lay an egg. The resulting pollen mass is all the food needed to support the developing larva. The process is repeated until the entire nest is fully outfitted. One or more generations are produced each year, depending on climate.

Believe It or Not: While many of the metallic green sweat bees nest in the ground and are solitary, it is not uncommon for individual females to share the same nest entrance with other females. This behavior is thought to help reduce nest parasitism by other bees.

Wasps, Bees, Ants, and Sawflies (Order Hymenoptera)

On Flowers

Yellow Jacket

Size: 0.4–0.6 inch long
ID Tips: Stout body with yellow-and-black markings and narrow, amber wings
Range: Throughout the United States

Yellow jackets are social wasps that are regularly mistaken for bees. Their paperlike nests are made from chewed wood and saliva and built in protected cavities, such as old underground mammal burrows, tree cavities, or hollow logs. However, they may also frequently nest in or around buildings. Yellow jackets aggressively defend their nests and will attack with the slightest provocation if disturbed. They can inflict a painful sting, so be cautious if you spot one, and don't approach an active nest. A single colony will grow continuously throughout the summer and may contain several thousand occupants. To support their developing young, adult yellow jackets are expert hunters and diligently collect a wide assortment of insect prey, including many garden pests. As fall approaches, the colony dies, but mated queens overwinter until the following spring when they start a new nest.

Believe It or Not: Adult yellow jackets are fond of sugary food and investigate open soda cans, beer bottles, or cut fruit with regularity. As a result, they may be a minor nuisance at picnics or other outdoor gatherings with food present.

Wasps, Bees, Ants, and Sawflies (Order Hymenoptera)

On Flowers

European Honey Bee

Size: 0.4–0.75 inch long

ID Tips: Fuzzy appearance, large black eyes, two pairs of wings, black-and-golden-orange-striped abdomen

Range: Throughout the United States

Small but mighty, the European Honey Bee is critical for global food production. Adult bees pollinate many valuable crops, including everything from blueberries and watermelons to apples and almonds. As a result, they are responsible for about one out of every three bites of food we eat each day. Honey bees are social insects. They live and work together in large colonies called hives, which may contain up to 100,000 individuals. Each hive has a single queen that is only responsible for egg laying. Most of the remaining bees are workers. Among other tasks, they visit flowers to collect pollen and nectar, the basic resources needed to produce food for the hive. The nectar is made into honey. This high-energy diet is stored in structures called honeycombs and eaten by bees during the winter when flower nectar is unavailable. Warning! While they are not typically aggressive, honey bees can inflict a painful sting, and this can lead to an allergic reaction in some people. Avoid disturbing bees or handling them, and stay far away from any active hives.

Believe It or Not: Queen bees can lay up to 1,500 eggs each day in the hive.

Wasps, Bees, Ants, and Sawflies (Order Hymenoptera)

Thread-waisted Wasp

Size: 0.4–1.2 inches long

ID Tips: Body generally black, often colored with red or yellow; abdomen very long and with a noticeably elongated waist

Range: Throughout the United States

Found across much of North America, these very distinctive-looking wasps are named for their elongated and stalked abdomens that give them the appearance of having narrow, threadlike waists. They are solitary wasps that typically construct nests in the ground. Other group members, such as mud daubers, build nests on structures out of moist soil. The adults feed on flower nectar and are also active predators, preying on butterfly and moth caterpillars, grasshoppers, and other sizable insects. Once targeted, the wasp grabs the victim with its mandibles, paralyzes it with a sting, and transports the insect back to the underground nest. The available food item is then devoured by the developing wasp larvae.

Believe It or Not: Like many other predatory insects, thread-waisted wasps are highly beneficial insects that often prey on many garden and agricultural pest insects. They are also considered important pollinators.

Wasps, Bees, Ants, and Sawflies (Order Hymenoptera)

On Flowers

Bald-faced Hornet

Size: 0.5–0.75 inch long

ID Tips: Stout black body with cream markings and narrow, transparent wings

Range: Throughout the United States

Despite its name, this large wasp is actually a species of yellow jacket, not a true hornet. It is similar in appearance to other yellow jackets but lacks the traditional bold bee-like yellow-and-black pattern. Instead, the bald-faced hornet is primarily black with some whitish markings and a distinctive white face. Additionally, they construct gray papery egg-shaped nests in trees or tall shrubs. While most are located just a few feet off the ground, some may be as high as 50 or more feet in the canopy. As nests are often concealed among vegetation, they may go unseen until late fall when branches are bare. Their hidden nature also makes it possible to accidentally disturb a nest. Like other social wasps, bald-faced hornets will aggressively defend their nests, with such interactions often resulting in painful stings. Be cautious and never closely approach or aggravate an active colony. Adults feed on nectar at flowers and also prey on other insects to feed their developing brood.

Believe It or Not: In addition to repeatedly stinging in defense, bald-faced hornets are also able to spray venom at potential predators. This behavior can be particularly effective if aimed at the eyes of nest intruders.

Wasps, Bees, Ants, and Sawflies (Order Hymenoptera)

Carpenter Bee

Size: 0.5–1.0 inch long
ID Tips: Robust shiny black head and abdomen; dense yellow hairs on the thorax
Range: Throughout the United States

These black-and-yellow solitary insects are some of the largest bees in North America. They resemble bumblebees but have a noticeably bare and shiny abdomen. Carpenter bees nest in wood and use their strong mouthparts to excavate round tunnels in dead trees, fence posts, or even houses and barns, producing a loud chewing or buzzing noise in the process. They prefer bare untreated wood and regularly reuse old nests. Within each tunnel, the female bee produces linear rows of individual brood cells. She then actively forages in the surrounding landscape to collect flower pollen and nectar, blends the two substances together to create a food ball, places one ball in each cell, and lays an egg. When complete, she seals off the cell and begins collecting floral resources again until all the cells are stocked. This bounty provides the resulting larvae with the food they need to develop. Adult carpenter bees overwinter in these wooden tunnels and mate the following spring before establishing new nests.

Believe It or Not: Despite their formidable appearance, carpenter bees are rarely aggressive. In fact, male carpenter bees can't even sting.

Wasps, Bees, Ants, and Sawflies (Order Hymenoptera)

On Flowers

Bumblebee

Size: 0.75–1.5 inches long

ID Tips: A generally fuzzy appearance; prominent black-and-yellow pattern

Range: Throughout the United States

Bumblebees are large, common visitors to flowers. Easy to recognize because of their fuzzy yellow-and-black body, they are highly beneficial and important pollinators. Most are social, producing one generation each year as part of a larger colony. Mated queen bumblebees overwinter and emerge in early spring to start a new nest, usually in or near the ground. As the colony expands, worker bees take over the primary duties of collecting pollen and nectar to feed the growing larvae. When you spot a bumblebee or its nest, be cautious. They actively defend their nest if disturbed and can deliver a very painful sting. In fact, their bold patterns warn predators of their potent defensive ability. As fall approaches, the colony begins to produce potential new queens and males. These reproductive bees leave the nest to mate. Soon the remaining colony will die. Only the new, young queens overwinter to start the cycle again the following year.

Believe It or Not: Unlike many other insects, bumblebees can regulate their body temperature by rapidly contracting their flight muscles—essentially shivering. This enables them to fly and actively forage for food in cold weather.

Wasps, Bees, Ants, and Sawflies (Order Hymenoptera)

Cicada Killer

Size: 1.1–1.9 inches long
ID Tips: Large; reddish-brown head and thorax, amber wings, and black abdomen marked with elongated yellow spots
Range: From the Rocky Mountains east

Without a doubt, cicada killers are impressive, formidable-looking wasps. Widely distributed across North America, the adults feed on flower nectar and are commonly encountered at blossoms where their large size quickly distinguishes them from most other bees and wasps. Females dig extensive underground burrows, each with several brood cells, in loose, well-drained soils. Once the burrows are complete, the female turns her attention to hunting. As their name implies, female wasps scour trees and branches for cicadas. When found, the wasp stings the prey to paralyze it, turns the cicada on its back, and transports it back to the burrow. Each cell is outfitted with one or more cicada. When adequately supplied, the female lays an egg and seals off the cell. The developing larva will feed to maturity on the well-stocked food reserves before emerging as a new adult wasp the following year.

Believe It or Not: Although not aggressive, female cicada killers can deliver a very painful sting if handled or provoked, so be careful. Males do not sting.

Spiders (Order Araneae)

On Flowers

Flower Crab Spider

Size: 0.12–0.5 inch long

ID Tips: Color highly variable; flattened body with rounded abdomen; the front two pairs of legs are much longer than hind legs; females are much larger than males

Range: Throughout the United States

This fascinating group is well known for a tremendous diversity of colors, which often enable the spiders to seamlessly blend in to their surroundings and avoid detection. They are ambush predators and don't use silk webs to capture their prey. Instead, they sit quietly on flowers or other vegetation frequented by mobile insects. Once in range, the flower crab spider quickly pounces on the unsuspecting victim and paralyzes it with venom. They possess particularly potent venom and thus are able to capture sizable, robust insects, including bees, butterflies, and true bugs, even those that are much larger than themselves. The bodies of flower crab spiders appear somewhat disproportionate. Their two front pairs of legs are much longer and held outward compared to those in the rear, making them resemble miniature crabs.

Believe It or Not: Some flower crab spiders can change their body color to better match the flower on which they're waiting to ambush prey.

Spiders (Order Araneae)

Green Lynx Spider

Size: 0.5–0.85 inch long
ID Tips: Bright green body; somewhat paler legs bearing black spines and spots
Range: Throughout the southern United States

The green lynx spider is a beautiful, emerald-green predator with catlike skills and acute vision. Frequently encountered on flowers or poised on the top of weedy plants blending seamlessly into the surrounding vegetation, it patiently sits in wait for insects to venture nearby. Once in sight, the spider cautiously approaches and then swiftly pounces on its unsuspecting prey, quickly immobilizing the insect with its venomous bite. The green lynx spider is more than capable of subduing insects larger than itself, including bees, butterflies, and even caterpillars. Like other spiders, females enclose their eggs in a silken sac. Once constructed, the green lynx spider aggressively guards her egg sac as well as her newly hatched young.

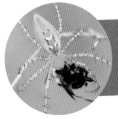

Believe It or Not: Adult spiders are capable of spitting venom when threatened. Using it like a natural pepper spray, they are the only known spider to employ this unique defensive strategy, though other spiders spit to capture prey.

On Flowers

Butterflies and Moths (Order Lepidoptera)

Plume Moth

Size: Wingspan 0.5–1.5 inches

ID Tips: Small; elongated slender body and dull-colored narrow wings that resemble a T shape when at rest

Range: Throughout the United States and Canada

This is a family of primarily small, drab moths; more than 160 species of plume moths are found throughout the United States and Canada. The adults are quite distinctive with narrow wings that are deeply divided into fringed lobes, much like the feathers of a bird. This characteristic is the basis for this group's name. The wings are held tightly together at a 90-degree angle from the body while at rest, giving the moth a very prominent T-shaped posture. At first glance, they resemble a small dead twig and look nothing like a tasty meal worthy of a predator's attention, and this camouflage has helped them survive. Plume moths are active during the day and commonly found at flowers where they feed on nectar. They can also be encountered at artificial lights at night, where their appearance clearly separates them from any other backyard moth. The larvae of some species are pests of certain crops or ornamental plants.

Believe It or Not: While some plume moths are considered pests, others are quite beneficial and have been used to help control various invasive plant species.

Butterflies and Moths (Order Lepidoptera)

On Flowers

Eastern Tailed Blue

Size: Wingspan 0.75–1.0 inch

ID Tips: Small; male bright blue above with narrow brown border; female brown above; both sexes have one or two small orange hind wing spots and a single, hairlike tail at the bottom of both wings

Range: Much of the United States, particularly in the East

This diminutive species is one of the most widespread and abundant butterflies in the eastern United States. Named for its distinctive hind wing tails, the eastern tailed blue can be quickly separated from virtually all other similar small blue butterflies based on this single characteristic. Adults have a weak, waltzing flight and erratically maneuver amongst low vegetation. They are extremely fond of flowers and easily attracted to gardens.

Unlike many other similar hairstreaks or blues, the eastern tailed blue often perches with its wings open. Males often gather in small "puddle clubs" at damp sand or gravel. The small sluglike larvae feed on the blossoms of a wide variety of primarily weedy plants in the bean family, including clovers and vetches.

Believe It or Not: Closely resembling antennae, the eastern tailed blue thin tails help deflect small predators—such as jumping spiders—away from the insect's vulnerable body.

Butterflies and Moths (Order Lepidoptera)

On Flowers

Reakirt's Blue

Size: Wingspan 0.75–1.0 inch

ID Tips: Small; wings above violet blue with white fringes; wings below light gray with numerous white and black marks, including a prominent row of large black, white-rimmed spots on the forewing

Range: The central and western United States

This is a common small blue butterfly of open sunny habitats across much of the western and central United States. A year-round resident from Central America to the desert Southwest and southern Texas, it regularly expands its range northward each year to temporarily colonize locations throughout the Great Plains. Despite its weak flight, adults are capable of traveling long distances. As a result, Reakirt's blue has been recorded as far to the east as Ohio and northward to the Upper Great Lakes and the Canadian border. The sluglike larvae feed on the buds and developing flowers of various pea family plants. Numerous generations are produced each year, depending on latitude. The butterfly can be found throughout the year in extreme southern locations.

Believe It or Not: Like many members of the family, Reakirt's blue larvae are often tended by ants. This unique association is considered mutually beneficial. Ants receive a sugary food reward from specialized organs on the larvae. In return, the ants act as bodyguards, aggressively defending the larvae from attacking insects or other small predators.

Butterflies and Moths (Order Lepidoptera)

Ailanthus Webworm Moth

Size: Wingspan 0.75–1.2 inches

ID Tips: Orange forewings with bands of cream spots edged in black; long antennae; wings held rolled over its back

Range: The eastern United States and Canada

At first glance, this unusual insect does not resemble a moth at all. The adults hold their narrow brightly colored wings tightly rolled over their backs, giving them an elongated tubelike shape more akin to a beetle. They are active during the day and are readily seen at flowers in open fields and meadows. The moth's distinctive name comes from its primary host, Ailanthus or Tree of Heaven, an exotic weedy tree native to Asia that has aggressively spread across much of the United States and southern Canada. The moth is also an invader, originating from Central and South America, where it feeds on a closely related species. Able to thrive on Tree of Heaven, the Ailanthus webworm moth has spread with the tree across much of eastern North America. The larvae feed gregariously on the host and construct nests by pulling together multiple leaves with loose silk webbing. While multiple generations are produced each year, it is unlikely that the moth can survive the harsh winters in the northern portion of its range.

Believe It or Not: The Ailanthus webworm moth is considered a good pollinator. Look for them on a number of small-flowered blooming plants, such as milkweed and goldenrod.

Butterflies and Moths (Order Lepidoptera)

On Flowers

Common Checkered/ White Checkered Skipper

Size: Wingspan 0.75–1.25 inches

ID Tips: Small; wings black with numerous white spots giving an overall checkered pattern

Range: Throughout the United States

Living up to its name, the common checkered skipper is widespread and generally abundant throughout much of the southern and central United States. It tends to expand its range northward each year, temporarily colonizing many northern states where it is most frequently encountered in late summer and early fall. Its occurrence as a seasonal vagrant at northern latitudes varies tremendously from year to year, however. Adults have a quick, erratic flight but regularly alight on low vegetation with their wings open. Nervous and combative males seem to continuously engage one another in a frenzy of activity. The green-gray larvae construct individual shelters on the upper portion of host plants by weaving several leaves together with silk. The small shelters help protect and conceal the larvae from predators. It can be found in most open, disturbed, and weedy sites.

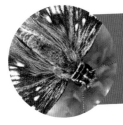

Believe It or Not: In the Deep South, the common checkered skipper has been almost completely replaced by the white checkered skipper. Virtually identical in both appearance and behavior, dissecting them is the only way to tell the two sibling species apart.

Butterflies and Moths (Order Lepidoptera)

Spring Azure

Size: Wingspan 0.75–1.25 inches
ID Tips: Wings below dusky gray with black spots and varying degrees of dark scaling; wings above pale blue with dark margins
Range: Throughout the United States

The spring azure is by far the most abundant and noticeable early-season butterfly in many parts of its range, often seen before many of the colorful spring flowering trees and shrubs are in bloom. A butterfly of woodlands and associated trails and margins, it periodically wanders into nearby open areas, including fields, suburban yards, meadows, and parks. Adults have a moderately slow flight and erratically scurry from ground level to canopy height, moving just over the surface of the vegetation. Males often congregate at mud puddles or on damp gravel roads in good numbers. They are extremely fond of flowers and readily land to sip available nectar. Adults feed and rest with their wings closed, so the lovely light blue color of their wings above can only be seen in flight or occasionally when basking. The slug-like larvae feed on the buds and developing flowers of various trees and shrubs.

Believe It or Not: While most commonly seen during the day, spring azures have occasionally been reported to be attracted to artificial lights at night, an unusual behavior for butterflies.

Butterflies and Moths (Order Lepidoptera)

On Flowers

Summer Azure

Size: Wingspan 0.8–1.3 inches

ID Tips: Small; wings above uniform pale blue with narrow dark margins in male; female is lighter blue with broad dark forewing margins; wings below chalky white with small dark spots and bands

Range: From the Great Plains east and into southern Canada

Although now considered a distinct species, the summer azure was previously viewed as a lighter, second-generation form of the early season spring azure. It is widespread and common throughout the eastern United States and southern Canada. Adults of this small dusty blue butterfly are found in and along woodlands, but they readily venture out into nearby open areas in search of nectar and may frequently wander into suburban yards and gardens. They have a moderately slow, dancing flight and unlike most other blues are often seen fluttering high among the branches of trees and shrubs. Females deposit their tiny whitish-green eggs on the buds of various woody plants. The resulting sluglike larvae feed on the developing flowers. Males often congregate in numbers at moist earth and mud puddles.

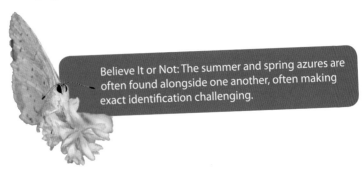

Believe It or Not: The summer and spring azures are often found alongside one another, often making exact identification challenging.

Butterflies and Moths (Order Lepidoptera)

European Skipper

Size: Wingspan 0.90–1.10 inches

ID Tips: Wings above bronzy orange with dark borders; wings below uniform unmarked orange

Range: Much of the northern United States and southern Canada

This diminutive bright-orange skipper is a regular sight in open, grassy areas throughout southern Canada and the northern half of the United States. Quite fond of flowers, the European skipper is a common visitor to gardens, parks, and yards with abundant blossoms and adapts well to more urban environments. The elongated pale green larvae feed on a wide range of different grasses. Adults maneuver close to the ground with a slow and somewhat erratic flight, typically unlike the frenetic behavior of many "skippers." Like many other grass skippers, adults feed and rest with their wings spread in a characteristic posture where the hind wings are lower than the forewings. This gives the small skipper the appearance of a fighter jet ready for takeoff.

Believe It or Not: As its name implies, the species was accidentally introduced into Ontario, Canada, from Europe in 1910 and continues to expand its range.

Butterflies and Moths (Order Lepidoptera)

Coral Hairstreak

Size: Wingspan 0.9–1.25 inches
ID Tips: Wings below light gray brown with a row of bright coral-colored spots along the outer margin of the hind wing; tailless
Range: Throughout much of the United States

This distinctive tailless hairstreak is unlikely to be confused with any other small butterfly. The coral hairstreak frequents a variety of semi-open, brushy habitats in close association with their larval host plants, which often grow aggressively and form thickets. Adults have a quick, erratic flight and readily perch on the tops of small trees or shrubs. Both sexes avidly visit flowers and are particularly fond of milkweed blossoms, where they are often observed feeding alongside other hairstreaks. The females deposit their small eggs on host branches. They overwinter and hatch the following spring. Only one generation is produced each year.

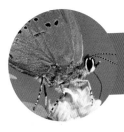

Believe It or Not: The sluglike larvae hide during the daytime in leaf litter below the host plant, coming out under the cover of night to feed.

Butterflies and Moths (Order Lepidoptera)

American Copper

Size: Wingspan 0.9–1.4 inches
ID Tips: Small; bright reddish-orange forewings with scattered black spots and gray hind wings with broad scalloped orange border
Range: Northern two-thirds of the United States

Although called the American Copper, some suggest that the eastern populations of this butterfly may actually be the result of historical introductions from Europe. This argument is fueled by the species' unique preference for open, disturbed habitats and primary use of weedy, non-native larval hosts—sheep sorrel and curry dock. It tends to occur in widespread and localized colonies but is often common to abundant when found. The bright reddish-orange adults frequently perch on low vegetation or bare soil with their wings held in a characteristic, partially open position. They have a somewhat weak, erratic flight and are particularly easy to closely observe when feeding. When you first see one in bright sunlight, you'll have no doubt why this butterfly is called a copper.

Believe it or not: While often somewhat spotty in occurrence, the American copper can at times be widely abundant. It is not uncommon for hundreds of individuals or more to be found in the right location.

103

Butterflies and Moths (Order Lepidoptera)

On Flowers

Black and Yellow Lichen Moth

Size: Wingspan 1.0–1.25 inches

ID Tips: Elongated black wings with orange bases; an iridescent blue-black body

Range: From the Rockies east in the United States and Canada

This boldly colored species is common and widespread throughout much of the United States and Canada east of the Rocky Mountains. Adults are active during the day and readily visit a wide range of small-flowered blossoms for nectar. As a result, they are considered beneficial pollinators. The sparsely haired larvae feed on lichens and sequester defensive compounds from their hosts. These are passed on to adult moths during development and help provide protection from potential predators. The stark orange-and-black wing pattern helps advertise their unpalatability and is shared by other members of the family, as well as some net-winged beetle species. This shared color pattern presumably reinforces the visual signal and encourages predators to stay away.

Believe It or Not: Lichen is sensitive to air pollution, particularly in more urban locations. As a result, lichen moths may also serve as valuable indicators to help monitor air quality and overall environmental health.

Butterflies and Moths (Order Lepidoptera)

On Flowers

Fiery Skipper

Size: Wingspan 1.0–1.25 inches

ID Tips: Elongated wings golden orange above with jagged black border; wings below golden orange with tiny scattered dark spots

Range: Primarily in the southern United States

The fiery skipper has a rapid, darting flight but frequently stops to perch on low vegetation. Adults are exceedingly fond of flowers and readily congregate at available blossoms, often forming little clouds of activity. They have a strong preference for colorful composites. It is a common to abundant skipper in gardens; suburban and urban yards; and open, weedy habitats across the southern United States. The fiery skipper readily expands it range northward each year to temporarily colonize parts of New England and the Great Lakes. The compact brownish larvae feed on a wide range of grasses, including those commonly used for lawns and golf courses. Numerous generations are produced each year.

Believe It or Not: The fiery skipper continues to expand its range northward due to climate change.

Butterflies and Moths (Order Lepidoptera)

On Flowers

Whirlabout

Size: Wingspan 1.0–1.25 inches

ID Tips: Elongated wings are golden orange above with black borders and a central black forewing spot in males and dark brown with cream spots in females; wings below are yellow in males to bronze in females with two bands of large dark spots

Range: The southeastern United States

The whirlabout is a diminutive skipper with two distinct rows of squarish spots on the hind wing below. This species is sexually dimorphic. Males are tawny orange and considerably brighter than their drab, brown female counterparts. Found abundantly across the Southeast, it regularly expands its range each summer, establishing temporary breeding colonies from Maryland to Texas. It is a butterfly of open, weedy sites, and the adults are quite fond of flowers and readily pause to feed at available blossoms. Where found, the whirlabout is commonly seen alongside the similar-looking fiery skipper. The chubby brown larvae feed on a variety of grasses, including those used for lawns. As a result, they can often be abundantly encountered along roadsides, in open fields, on forest margins, as well as in parks, yards, and gardens.

Believe It or Not: The name whirlabout is most fitting for this species, as adults frenetically flutter low over vegetation in seemingly restless waves of activity.

Butterflies and Moths (Order Lepidoptera)

Eight-spotted Forester

Size: Wingspan 1.0–1.5 inches

ID Tips: Wings blackish with yellow and white spots; bright orange color on front and middle legs

Range: Primarily the eastern United States; scattered westward

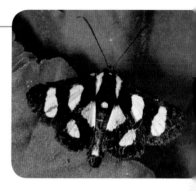

The eight-spotted forester is an energetic little day-flying moth and common across much of the eastern United States. As its name suggests, the somewhat shiny blackish-blue wings are adorned with eight large yellow-to-white spots, making identification easy even from some distance. From the side, this lovely moth reveals prominent bright orange tufts on it first two pairs of legs and a very fuzzy body. The adults are avid flower visitors and readily scurry from blossom to blossom to sip nectar. Because of this behavior and its overall colorful appearance, the eight-spotted forester is often mistaken for a butterfly. The white and orange caterpillars are covered with small black spots and feed on wild grape and Virginia creeper, two very abundant and somewhat weedy vines. One generation is produced in the north with two flights being the norm in more southern latitudes.

Believe It or Not: The eight-spotted forester overwinters in the pupal stage and can remain in a state of diapause, or dormancy, for several years.

Butterflies and Moths (Order Lepidoptera)

On Flowers

Gray Hairstreak

Size: Wingspan 1.0–1.5 inches

ID Tips: Small; slate gray above with distinct orange-capped black hind wing spot near the hind wing tail; wings below are light gray with white-and-black dashed bands and an orange-capped black spot near the hind wing tail

Range: Throughout the United States

The gray hairstreak may be found in just about any open habitat because it is able to utilize an enormous variety of plants as larval hosts. As a result, it may be the most frequently encountered hairstreak butterfly in North America, provided that you look closely and don't overlook the butterfly due to its small size. Adults have a quick, erratic flight and are exceedingly fond of flowers, making them common garden visitors. They regularly perch with their wings partially open, an unusual behavior for most hairstreaks. The small tails on the hind wing resemble antennae and help draw the attention of would-be predators away from the insect's vulnerable body. This deception, employed by many members of this butterfly family, is enhanced by the bright orange spots, and a characteristic wing-waggling behavior draws attention to this feature, known as a "false head."

Believe It or Not: Gray hairstreak larvae have an incredibly broad range of hosts, feeding on more than 200 different plant species in many different families.

Butterflies and Moths (Order Lepidoptera)

On Flowers

Sara Orangetip

Size: Wingspan 1.0–1.5 inches

ID Tips: Small; wings white to yellowish-white with a large orange patch in black that starts near the tip and runs to the forewing; the hind wing below has greenish marbling

Range: West Coast of the United States

Also called the Pacific orangetip, this attractive, small butterfly ranges along the West Coast from Alaska to Mexico. It is aptly named for the prominent reddish-orange tip on each forewing that is readily prominent, even from some distance away. While females are generally whitish, some can be quite yellow in color. A denizen of open habitats in the foothills and mountains, the Sara orangetip can be found in canyons, riparian areas, open woodlands, fields, roadsides, and hillsides. Although generally common, it can be quite localized in abundance. Adults have a quick, somewhat bouncing flight but readily pause at flowers where they can often be closely observed. The slender larvae feed primarily on the buds, flowers, and fruit of various plants in the mustard family.

Believe It or Not: It is often one of the earliest butterflies on the wing in the northern parts of its range.

Butterflies and Moths (Order Lepidoptera)

On Flowers

Bella Moth

Size: Wingspan 1.0–1.7 inches

ID Tips: Forewings yellow with white bands containing black spots; hind wings pink with black margins

Range: Primarily the eastern United States

Also called the ornate bella moth, this attractive insect is common and extremely widespread, occurring across much of the eastern United States west to Arizona and south through Mexico, Central and South America, and much of the Caribbean. The colorful adults fly during the day and are frequently encountered at flowers. The orange-and-black banded larvae feed on various legumes but prefer plants belonging to the weedy and often invasive genus Crotalaria, which is known as rattlebox because of its large, inflated dark seedpods that make a rattling sound when shaken. These plants are rich in noxious chemicals, which the developing larvae incorporate into their bodies while feeding. This renders them and the resulting adults toxic to various predators and also gives the moth its gaudy coloration.

Believe It or Not: The chemically protected moths are well defended against spiders, which are common predators of day-flying and flower-visiting insects. A variety of these predatory spiders actively reject adult bella moths.

Butterflies and Moths (Order Lepidoptera)

Great Purple Hairstreak

Size: Wingspan 1.0–1.7 inches

ID Tips: Wings above metallic blue; wings below black with red markings; hind wing with long hairlike tails

Range: The southern half of the United States

The great purple hairstreak significantly dwarfs most other members of its family. The upper wing surfaces of this stunning hairstreak are actually bright metallic blue, not purple, as its name suggests. Primarily a butterfly of forested areas, it usually dwells high in the canopy of trees. Their sluglike green larvae feed on mistletoe, a parasitic plant found on various trees. Despite its arboreal habits, adults regularly venture down to available blossoms where they can be closely observed. Quickly identified by its orange abdomen, the butterfly's bright and contrasting colors warn predators of its distasteful nature. It typically feeds and rests with its wings firmly closed, so the brilliant blue scaling is only seen in flight.

Believe It or Not: The base of the butterfly's long hind wing tails flair outward and resemble eyes; like other butterflies, it makes use of this "false head" disguise to trick potential predators.

Butterflies and Moths (Order Lepidoptera)

On Flowers

Pearl Crescent

Size: Wingspan 1.25–1.6 inches
ID Tips: Small; wings above orange with dark bands, spots, and borders
Range: From the Rocky Mountains east

This is one of the most abundant and readily encountered crescent butterflies in the eastern United States. At home in most open, sunny landscapes, it frequents old fields, meadows, and rural pastures, as well as gardens and urban parks. Its appearance varies seasonally; spring and fall adults are darker and more heavily patterned on the wings below compared to summer individuals. Adults have a rapid, erratic flight but readily pause to perch on low vegetation and frequently visit flowers. Females deposit the small, round eggs in clusters on host leaves. The small, brown caterpillars often feed together, especially when young.

Believe It or Not: The pearl crescent is an opportunistic breeder, producing new generations as long as favorable conditions allow.

Butterflies and Moths (Order Lepidoptera)

Checkered White

Size: Wingspan 1.25–2.0 inches
ID Tips: Wings white with charcoal spots and markings
Range: Throughout the United States

This is a common butterfly of open, often disturbed sites across the United States. Their larvae feed on a variety of weedy plants belonging to the mustard family. It can at times be abundant along roadsides or fallow agricultural lands and is typically found more often in rural areas. The checkered white is a resident of the southern states but can colonize north to the Canadian border, with few butterflies one year and many the next. As a result, populations often fluctuate considerably from year to year at more northern latitudes. Adults have a quick, somewhat erratic flight and can be a challenge to closely approach. Two or more generations are produced each year. The sexes differ dramatically in the amount of dark scaling on the wings. The more heavily patterned females can generally be told apart from the whiter, more immaculate-looking male butterflies, even from a distance.

Believe It or Not: While the bold checkered pattern is obvious to people, male and female butterflies recognize one another by ultraviolet light.

Butterflies and Moths (Order Lepidoptera)

On Flowers

Sleepy Orange

Size: Wingspan 1.25–2.0 inches

ID Tips: Wings above are bright orange with irregular black borders; wings below are variable, butter yellow to reddish-brown

Range: Primarily the southern half of the United States

Much debate centers on the origin of this butterfly's common name. One interpretation points to its narrow black forewing spot, which resembles partially closed eyes, as if the butterfly was tired. Others suggest that it gets its name from its rather lackadaisical behavior. Individuals in the far South overwinter as adults in reproductive diapause. They are highly sedentary during the winter but often become active on mild days to obtain nectar at available flowers before returning to their apparent slumber. In spring, the sleepy orange starts to breed again and colonize locations as far north as the Midwest each year. The adults have a low, fluttering flight and are fond of flowers. They can be common garden visitors.

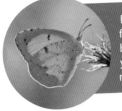

Believe It or Not: The adults produce distinct seasonal forms, which vary dramatically in color on the wings below. Individuals found in summer or spring are yellow; winter forms are a deep reddish-brown and more heavily patterned.

Butterflies and Moths (Order Lepidoptera)

Variable Checkerspot

Size: Wingspan 1.3–2.25 inches

ID Tips: Wings above are black with extensive white and reddish-orange spots; wings below are orange with black-outlined white spots

Range: The western United States

The variable checkerspot is a small attractive butterfly commonly found throughout western North America from Alaska to Baja California. Named for its characteristic and highly variable checkered pattern, some 30 different subspecies or geographic variants are recognized. It may be frequently encountered in a wide variety of habitats, including meadows, open woodlands, canyons, sagebrush hills, grasslands, and alpine tundra. Adults have a quick, darting flight but regularly pause to take nectar at available flowers or perch on rocks or low-growing vegetation, typically with wings spread. Not overly wary, adults are often easy to closely approach. Females deposit the yellow eggs in clusters on host plant leaves. The resulting spiny larvae feed gregariously in a silken nest when young, and as solitary individuals later in life. Depending on latitude and elevation, there can be one or more generations.

Believe It or Not: Like other boldly colored insects, the variable checkerspot is chemically protected and unpalatable to various predators.

Butterflies and Moths (Order Lepidoptera)

On Flowers

Cabbage White

Size: Wingspan 1.5–2.0 inches
ID Tips: White above with black forewing tips and black spots
Range: Throughout the United States

Accidentally introduced from Europe in 1860, the cabbage white or European cabbage butterfly quickly expanded across much of North America. Readily encountered in almost any open, disturbed area, it may be particularly noticeable in home vegetable gardens or farmers' fields, where it uses a variety of larval hosts in the mustard family, including cabbages, broccoli, and turnips. The gray-green caterpillars feed on the leaves and developing flower heads. Adults have a slow, somewhat awkward flight and often rest with their wings partially open, making it easy to tell the sexes apart. Female cabbage white butterflies have two black forewing spots, whereas males have one.

Believe It or Not: The cabbage white is one of the few butterfly species considered to be a serious agricultural and garden pest.

Butterflies and Moths (Order Lepidoptera)

Long-tailed Skipper

Size: Wingspan 1.5–2.0 inches
ID Tips: Wings are brown with bluish-green scaling on wing bases and body; hind wing has a long tail
Range: Primarily the southeastern United States

Resembling a small swallowtail, the long-tailed skipper is one of the most common and distinctive skipper butterflies in the Southeast. At home in most open, disturbed, and weedy sites, it regularly wanders into nearby yards and can be an abundant garden visitor. Adults have a quick, low flight but often stop to perch on low vegetation or obtain nectar at blossoms. The colorful larvae construct individual leaf shelters on the host by folding over small sections of a leaf with silk. Older larvae may use an entire leaf or connect several leaves together. They feed on a wide variety of weedy legumes, as well as cultivated garden beans. The adults and larvae are particularly abundant in late summer and fall.

Believe It or Not: The long-tailed skipper can't survive freezing temperatures. As a result, it is migratory and moves south each year to escape the cold. They can often be seen by the thousands moving into the Florida peninsula.

117

Butterflies and Moths (Order Lepidoptera)

On Flowers

Virginia Ctenucha

Size: Wingspan 1.5–2.0 inches
ID Tips: Metallic blue body with dull gray-black wings and an orange head
Range: Primarily the northeastern United States and the Upper Midwest

This unusual-looking creature is actually a day-flying moth and one of several species known to mimic wasps in appearance, which helps protect them from predators. The large adults have a somewhat weak, fluttering flight and frequently land on vegetation. They feed on flower nectar and may be quite common at various blossoms, particularly in open fields, meadows, or along forest margins. At first glance, the Virginia ctenucha may appear quite drab, but closer inspection, especially in bright sunlight, reveals a metallic blue body and a bright orange head. Pronounced "ten-OOCH-ah," the name is Greek for "has a comb" and refers to the moth's long ferny antennae. The hairy, compact caterpillars have two color forms, one dark with black and yellow hairs, and a noticeably lighter one dominated by white and yellow tufts. They overwinter as larvae and construct a loose cocoon the following spring.

Believe It or Not: The larvae of the Virginia ctenucha are most commonly seen wandering along the ground in late summer or early spring but may even be active on warm winter days.

Butterflies and Moths (Order Lepidoptera)

On Flowers

Hummingbird or Bumblebee Moth

Size: Wingspan 1.5–2.2 inches
ID Tips: Stout hairy body with partially transparent wings; golden thorax and a dark abdomen
Range: Throughout the United States

These entertaining day-flying moths are frequent visitors to yards and gardens that are rich with flowers. Like small hummingbirds, they adeptly maneuver from blossom to blossom, pausing briefly in front of each to uncurl their long proboscis and sip out sugar-rich nectar. In the process, they also help pollinate many of the flowers. The rapid motion of their narrow, semitransparent wings even causes a noticeable buzzing noise when the insects are near. Most hummingbird moths have a furry, banded gold appearance and resemble bumblebees. Like other members of the sphinx moth family, the larvae have a distinctive curved horn on their rear end and are often referred to as hornworms. When fully grown, they crawl off their host plant and pupate in the soil or leaf litter.

Believe It or Not: Adult moths are aerial masters and capable of flying backwards, hovering almost motionlessly in mid-air, and darting rapidly from one location to another.

Butterflies and Moths (Order Lepidoptera)

On Flowers

Common Buckeye

Size: Wingspan 1.5–2.25 inches

ID Tips: Brown hind wings with prominent eyespots; forewings have a broad white patch and two short orange bars

Range: Throughout much of the United States, especially the southern half

The common buckeye is one of our most distinctive butterflies. Widespread and frequently encountered across the southern half of the United States, it is intolerant of harsh winter temperatures. When the weather is warmer, however, it regularly colonizes many northern regions each year. The large, conspicuous target-shaped eyespots help deflect attacks from predators and serve to startle would-be predators. The butterfly is fond of most open, sunny locations and may be a frequent garden visitor, visiting a variety of low-growing flowers quite often. Adults regularly land on bare soil or gravel with their wings outstretched but are extremely wary and challenging to closely approach. Males readily establish territories and actively fly up to investigate most passing insects, only to return to the same area moments later.

Believe It or Not: The common buckeye survives the winter as an adult butterfly and produces distinct seasonal forms. Cool-season individuals have wings that are a rich reddish-brown color; summer forms are a lighter tan.

Butterflies and Moths (Order Lepidoptera)

American Snout

Size: Wingspan 1.6–1.9 inches

ID Tips: Wings are above brown with orange patches and white forewing spots; forewing tips are squared off

Range: Primarily the southern half of the United States; into Midwest and Northeast

The American snout gets its odd name from the long labial palpi (mustache-like paired mouthparts) that resemble an elongated, beaklike nose. This unique feature, combined with the mottled brown coloration underneath its wings, enhances the butterfly's overall "dead leaf" appearance when it is at rest. It is a permanent resident of the southern states but regularly moves north to temporarily colonize much of the United States each year. Adults have a quick, somewhat bouncy flight and frequently visit available flowers. They rest and feed with their wings tightly closed. Females lay the tiny white eggs in the leaf axils of host trees.

Believe It or Not: In the western United States, American snouts occasionally make mass northward migrations. Known as "outbreak years," these movements often involve millions of individuals. In 1922, 25 million American snouts *per minute* passed over a 250-mile front.

Butterflies and Moths (Order Lepidoptera)

On Flowers

Orange Sulphur

Size: Wingspan 1.6–2.4 inches

ID Tips: Wings above bright yellow-orange with black borders; wings below yellow with one or two central red-rimmed silvery spots

Range: Throughout the United States

This widespread North American species can be found in virtually any open landscape, including old fields, roadsides, yards, gardens, and parks. It may be most abundant in commercial clover or alfalfa fields, where it may occasionally become an economic pest. Because of this strong host preference, it is often also called the alfalfa butterfly. It is a prolific colonizer and readily populates new locations; it is commonly found alongside the similar-looking clouded sulphur. Adults have a rapid, somewhat erratic flight and scurry close to the ground over low vegetation. When resting or feeding, they always sit with their wings firmly closed. Both sexes are fond of flowers and may be regular garden visitors. Individuals produced in early spring are generally smaller and less intensely colored compared to summer adults.

Believe It or Not: Both the orange and clouded sulphur butterflies are dimorphic. Males are bright yellow-orange while females can be either yellow or all white.

Butterflies and Moths (Order Lepidoptera)

Painted Lady

Size: Wingspan 1.75–2.4 inches

ID Tips: Wings pinkish-orange above with dark markings and white spots near the tip of the forewing; wings below are brown with cream patches and an ornate cobweb pattern

Range: Throughout the United States

Unable to survive freezing temperatures, the painted lady typically overwinters in Mexico and annually recolonizes much of the North American continent each summer. As a result, it is often somewhat sporadic in many locations with populations varying considerably in abundance from year to year. It occasionally has huge population outbreaks. It can be found in just about any open, disturbed landscape where its weedy larval hosts abound. The adults have a rapid, erratic flight usually close to the ground, but they frequently stop to perch or visit flowers. The dark spiny larvae construct shelters of loose webbing on their host plants.

Believe It or Not: Painted lady butterflies are available commercially in rearing kits for home or classroom use. They are also widely used for celebratory event releases.

Butterflies and Moths (Order Lepidoptera)

On Flowers

Silver-spotted Skipper

Size: Wingspan 1.75–2.4 inches

ID Tips: Stout body; wings brown with distinct, elongated white patch in center of hind wing below

Range: Throughout the United States

This is one of the most abundant large skippers in the United States. Named for its distinctive, pure silver-white hind wing patch, the silver-spotted skipper frequents open, sunny habitats, including old fields, meadows, roadsides, gardens, yards, and parks. Adults have a powerful, darting flight that makes them challenging to pursue. Luckily, they are fond of flowers and readily pause to feed where they can be closely observed. They have a relatively long proboscis and can easily gain access to nectar from a wide range of blossoms. Males perch on shrubs or overhanging branches and aggressively investigate passing organisms. The plump larvae construct individual shelters on their host plants by tying one or more leaves together with silk.

Believe It or Not: Many predatory wasps locate potential prey, such as butterfly and moth caterpillars, by using chemical cues, including those given off by a caterpillar's waste products. As a result, silver-spotted skipper larvae forcibly fling poop out of their leaf shelters to avoid detection.

Butterflies and Moths (Order Lepidoptera)

Red Admiral

Size: Wingspan 1.75–2.5 inches
ID Tips: Wings dark brownish-black with a red hind wing margin and a red band through the middle of the forewing
Range: Throughout the United States

This medium-size butterfly is widespread and common throughout the United States. Its red-and-black pattern is distinctive and cannot be confused with any other butterfly. Typically unable to survive the winter in cold climates, it regularly recolonizes much of the northern latitudes each summer. As a result, its abundance often fluctuates from year to year and it may occasionally experience tremendous population outbreaks. Adults have a rapid, erratic flight but regularly perch on low vegetation or on the ground in sunlit areas. The red admiral frequents open woodlands, forest edges, sun-dappled yards, and adjacent open areas. The males establish and aggressively defend small territories, flying out to investigate passing butterflies and other large insects.

Believe It or Not: Adult red admirals occasionally visit flowers but more frequently feed on tree sap, animal dung, and rotting fruit.

Butterflies and Moths (Order Lepidoptera)

On Flowers

Cloudless Sulphur

Size: Wingspan 2.2–3.0 inches
ID Tips: Large; greenish-yellow wings with varying amounts of brownish markings
Range: The southern half of the United States

The cloudless sulphur is a large yellow butterfly with a fast, powerful flight. An abundant resident of the southern states, it regularly disperses northward to temporarily colonize more northern latitudes each year, occasionally reaching the Great Lakes and even the Canadian border. As fall approaches, the species undergoes a massive, southward migration. This annual migration is one of the Southeast's most impressive natural phenomena. Adults survive the winter and begin to disperse northward again as spring arrives. Adults have an extremely long proboscis and can feed at many long, tubular flowers that are typically inaccessible to other butterflies.

Believe It or Not: The sizable larvae have different color forms depending on which part of the host plant they feed. Larvae munching away on leaves will be predominantly green while those feeding on the showy blossoms will generally be yellow to match the background color of the flower.

Butterflies and Moths (Order Lepidoptera)

White-lined Sphinx Moth

Size: Wingspan 2.2–3.5 inches
ID Tips: Stout elongated body; elongated forewings are brown with a pale stripe; hind wings are pink with dark borders
Range: Throughout the United States

Like other sphinx moths, this showy insect is a master of flight. Its robust but streamlined body and long, narrow wings enable the adult to adeptly maneuver amid vegetation, hover effortlessly in midair, and rapidly dart from one location to another. Fond of flowers, it uncurls a long, thin proboscis to access sugary nectar in tubular blossoms, feeding much like a hummingbird, with which it is often confused. Just as with a hummingbird, powering the speedy wingbeats requires a lot of energy. As a result, white-lined sphinx moths preferentially visit flowers that provide high-quality foods. The primarily nocturnal adults rest during the day and become active at dusk, when they are most often spotted in gardens and yards. The colorful larvae are highly variable in color and pattern but all have a characteristic thornlike projection on the rear end. Most sphinx moth larvae share this feature, which is the inspiration for their common name of "hornworms."

Believe It or Not: White-lined sphinx moth larvae are periodically seen crawling along the ground in large numbers. These mass movements are likely triggered by a need to locate additional food or appropriate sites in which to pupate.

Butterflies and Moths (Order Lepidoptera)

On Flowers

Gulf Fritillary

Size: Wingspan 2.5–3.0 inches

ID Tips: Wings above are bright orange with black markings; wings below are brownish-orange with elongated silvery spots

Range: Primarily the southern United States

The underside of the gulf fritillary's hind wings are beautifully decorated with numerous silvery, mercury-like patches that quickly distinguish it from the equally common monarch butterfly. A resident of the extreme southern part of the country, it regularly expands it range northward each year to temporarily colonize much of the central United States. Exceptionally fond of flowers, it is a regular and often abundant garden visitor. Adults have a quick, somewhat low flight but regularly pause to feed at available blooms. The bright orange larvae have long-branched black spines and utilize various species of native and cultivated passionflower vines as hosts.

Believe It or Not: The gulf fritillary is one of several migratory butterfly species. Others include monarchs, cloudless sulphurs, long-tailed skippers, and common buckeyes. All move south each year in large numbers to escape freezing winter temperatures.

Butterflies and Moths (Order Lepidoptera)

Anise Swallowtail

Size: Wingspan 2.75–3.2 inches
ID Tips: Wings black with a broad yellow band; a single hind wing tail
Range: The western United States

The anise swallowtail is a common butterfly of western North America east to North Dakota and Nebraska. Frequenting a variety of open habitats, including old fields, meadows, and roadsides, it is also at home in more suburban locations and may be a regular garden visitor. Males often prefer hilltops and patrol them for passing females. The large green-and-black larvae feed on a number of wild and cultivated members of the carrot family, such as fennel, dill, parsley, and, of course, anise, after which it is named. Depending on latitude and elevation, one or more generations may be produced each year with more in warmer environments.

Believe It or Not: The anise swallowtail can be found in a wide variety of open habitats—from coastal gardens to high mountains.

Butterflies and Moths (Order Lepidoptera)

On Flowers

Great Spangled Fritillary

Size: Wingspan 2.9–3.8 inches

ID Tips: Large; wings above are bright orange with black markings; hind wings below are orange-brown with numerous large metallic silver spots

Range: Throughout much of the northern two-thirds of the United States

This large and conspicuous species is arguably one of the most common fritillaries in the United States. The bright orange adults have a strong, directed flight but regularly pause to take nectar at available flowers. In fact, it is not unusual to see multiple individuals at a single blossom. The great spangled fritillary is a butterfly of open, moist fields and meadows and roadsides that are often adjacent to woodlands. It has a single, protracted summer flight, with males often emerging several weeks before females. Females deposit their small eggs on or near host violets, often laying more than 1,000 during a lifetime. Newly hatched larvae overwinter and complete development the following spring.

Believe It or Not: Some nine different subspecies of this butterfly are known.

Butterflies and Moths (Order Lepidoptera)

Monarch Butterfly

Size: Wingspan 3.5–4.0 inches
ID Tips: Orange with black veins and wing borders
Range: Throughout the United States and southern Canada

The monarch is the most recognized butterfly in North America. During spring and summer, monarchs breed throughout the United States and southern Canada. In the fall, adults east of the Rocky Mountains migrate to Mexico, overwintering by the millions in high-elevation fir forests. In the western United States, monarchs migrate to scattered sites along the coast of California. The following spring, these butterflies leave their overwintering sites and fly northward in search of host plants on which to lay their eggs. Monarchs feed almost entirely on milkweed plants, which contain toxic chemicals called cardenolides. As the striped caterpillars feed, the chemicals from the plants accumulate in their body, making them and the resulting adult butterflies taste very bad to birds and other predators. The monarch's brilliant orange-and-black wing colors warn predators to stay away.

Believe It or Not: Some adult monarchs fly up to 3,000 miles during the fall migration.

Butterflies and Moths (Order Lepidoptera)

On Flowers

Eastern Tiger Swallowtail

Size: Wingspan 3.5–5.5 inches

ID Tips: Large; yellow wings with black stripes and black margins; each hind wing with a single tail; some females are all black

Range: The eastern United States

Living up to its name, the majestic eastern tiger swallowtail's bold yellow and black markings are distinctive. While males are always yellow, female butterflies are dimorphic—they're either yellow with black stripes or all black, which helps them mimic the toxic pipevine swallowtail. Although fond of woodlands and waterways, the eastern tiger swallowtail is equally at home in suburban yards and urban parks. Exceedingly fond of flowers, the adults seldom flutter their wings while feeding. They instead rest on the blossoms with their colorful wings outstretched. The sizable green larvae have prominent false eyespots on an enlarged thorax, giving them the appearance of a small lizard or snake. When not actively feeding, they sit on the upper surface of their host leaves, including ash, tulip tree, and wild cherry. Two or more generations are produced each year.

Believe It or Not: Some female tiger swallowtails can approach six inches in wingspan, making them the largest butterfly in eastern North America.

Butterflies and Moths (Order Lepidoptera)

Two-tailed Swallowtail

Size: Wingspan 3.5–6.0 inches
ID Tips: Large; yellow wings with bold black stripes and two hind wing tails
Range: The western half of the United States

This impressive insect is one of the largest butterflies in western North America. While similar in appearance to the eastern tiger swallowtail, it can be differentiated based on its two namesake hind wing tails, although the second is noticeably shorter than the first. Adults have a powerful, swift flight but readily come to flowers where they feed with their broad wings outstretched. The two-tailed swallowtail is commonly encountered in urban areas, roadsides, and parks where its larvae feed on ash and chokecherry trees. Older larvae have prominent false eyespots on the enlarged thorax; this helps them potentially startle predators.

Believe It or Not: When young, two-tailed swallowtail larvae are black and white and resemble bird droppings. This unique disguise is quite effective, as no right-minded bird would wish to eat its own excrement.

Bees, Wasps, and Ants (Order Hymenoptera)

On Structures

Mud Dauber Wasp

Size: 0.75–1.0 inch long
ID Tips: Variable; typically black to metallic bluish-black body, often with some yellow markings; a long, narrow waist
Range: Throughout the United States

This is an entertaining group of solitary wasps known for its architectural prowess. Most are medium-size, slender insects with long, thread-like waists. As their name suggests, mud daubers construct nests made of mud. Female wasps are the ones that do the work. They are industrious creatures that visit pond edges or other moist sites repeatedly to collect small balls of mud with their jaws. They then transport this raw material to a nest site and mold it into place until construction is complete. The specific design of the nest differs with each particular species. Once the nest is finished, the female will collect available small prey, including other insects and spiders, a particular favorite of theirs. She places them in individual cells, lays one egg per cell, and then seals up each cell with mud until all are fully partitioned. The paralyzed prey will serve as food for her developing larvae. As new wasps emerge, they start the process all over again. While mud daubers are considered beneficial insects, their nests are often considered a nuisance when constructed on homes or other buildings.

Believe It or Not: Not all mud daubers construct their own nests. Some, like the blue mud dauber, choose to use the abandoned nests of other mud dauber species.

Bees, Wasps, and Ants (Order Hymenoptera)

On Structures

Paper Wasp

Size: 0.75–1.0 inch long
ID Tips: Variable; slender brown to black bodies with yellow markings; narrow waist; pointed abdomen
Range: Throughout the United States

These social insects are named for their brown papery nests, which they construct by combining plant material with saliva. Having a characteristic open comb with numerous deep hexagonal cells to hold their developing young, the nests are common around homes and often found in sheltered sites, such as under eaves, overhangs, or around door frames. Be careful! While generally not aggressive unless disturbed, paper wasps can deliver very painful stings. It is therefore best to keep a safe distance away from active nests. The adult wasps are actually considered highly beneficial. They feed on nectar and thus help pollinate flowers. They are also tenacious hunters, capturing a wide variety of fleshy insects, including many garden pests. The resulting prey is carried back to the nest to feed the hungry wasp larvae.

Believe It or Not: Most paper wasp colonies support between 1-6 dozen individuals.

True Flies (Order Diptera)

On Structures

House Fly

Size: 0.15–0.30 inch long

ID Tips: Gray hairy body with four black stripes on the thorax; transparent wings and red eyes

Range: Throughout the United States

As its name suggests, this insect is likely the most commonly encountered fly in and around the home. Drab and dark, it is closely associated with the presence of humans and generally considered a nuisance pest, frequently landing on food. House flies are strongly attracted by garbage, animal waste, and decaying organic matter on which they may feed and lay eggs. Their elongated whitish larvae, called maggots, develop quickly especially in nutrient-rich materials. More than a dozen generations may be produced per year, especially in warm climates. Adult house flies do not bite. They use their spongelike mouthparts to sop up liquids and are especially fond of sugary foods.

Believe It or Not: House flies can feed on solid food items. However, they must release salivary secretions and regurgitated food to help initiate the digestive process. They then suck up the resulting liquefied materials.

Spiders (Order Araneae)

Cellar Spider

Size: Body is 0.25–0.35 inch long; much longer legs
ID Tips: Small oblong gray to brown body with eight very long thin legs
Range: Throughout the United States

Due to their small body size and thin elongated legs, these flimsy-looking spiders are often mistaken for daddy longlegs, a group of arachnids more closely related to scorpions than to spiders. By contrast, cellar spiders are true spiders and have a typical two-part body that consists of a cephalothorax and a distinct abdomen, eight eyes, and fanglike mouthparts. They are commonly found in homes or other structures and spin silk, creating what appears to be disheveled, highly unorganized small webs for capturing prey. They build their webs in shady corners, basements, attics, or under overhangs. They subsist on a variety of small insects but may also invade the webs of other spiders and eat the unsuspecting owner.

Believe It or Not: When disturbed, cellar spiders often vibrate rapidly in their webs. The resulting quick, blurring motion tends to confuse potential predators, making a direct attack more challenging.

Sowbugs, Pillbugs, and Woodlice (Order Isopoda)

On the Ground

Sowbug and Pillbug

Size: 0.25–0.5 inch long

ID Tips: Dark gray to brown; oval body with seven platelike segments and seven pairs of legs

Range: Throughout the United States

Where to Look: Under logs, rocks, leaf litter

Sowbugs and pillbugs are odd, primitive-looking creatures. Not insects at all, they are instead small, gray, land-dwelling crustaceans that are related to crayfish. Resembling tiny armadillos, they have hard, shelllike coverings made up of segmented plates and seven pairs of legs. While sowbugs and pillbugs are very similar in appearance, pillbugs can roll up into a ball when disturbed, earning them the name "roly-poly." Both are active at night, feeding on decaying plant material. This beneficial activity helps return organic nutrients to the soil. During the day, they can commonly be found under logs, rocks, leaf litter, or in other dark and damp locations. They are unable to tolerate dry environments and will die quickly without moisture.

Believe It or Not: Females carry their eggs in a sac on the underside of the body. After hatching, the young may remain with the mother for a few weeks before wandering away to fend for themselves.

Beetles (Order Coleoptera)

On the Ground

Ground Beetle

Size: 0.25–1.0 inch long
ID Tips: Brown to black; elongated, somewhat flattened body; large jaws; long legs
Range: Throughout the United States
Where to Look: Under logs, rocks, leaf litter

Ground beetles are active, nocturnal predators that use their sizable jaws to capture various insects and other soft-bodied invertebrates, including earthworms, slugs, and snails. While generally dark and drab in color, some species have shiny metallic hues and are quite attractive. During the day, look for them under logs, rocks, leaves, or other debris. Some species may also be attracted to artificial lights. Ground beetles have long, powerful legs and are capable of running quickly if disturbed, but they seldom fly. Adults even scurry up vegetation in search of potential prey. Their elongated larvae are also ferocious hunters and search for food below the soil's surface.

Believe It or Not: Ground beetles can defend themselves by emitting noxious chemicals from the tip of their abdomen. Arguably the most well-known example is the bombardier beetle, which can spray a jet of superheated fluid with pinpoint accuracy.

Beetles (Order Coleoptera)

On the Ground

Tiger Beetle

Size: 0.5–0.9 inch long

ID Tips: Shiny, often iridescent, patterned body; large eyes; long legs; prominent jaws

Range: Throughout the United States

Where to Look: On the ground

Like their namesake, tiger beetles are active and ferocious hunters. Typically found in a variety of open, sandy habitats, they are built for speed, and they hunt down their prey by running them down over open ground. Once in range, they seize insects or other invertebrates with their impressive jawlike mandibles, often subduing organisms larger than themselves. Tiger beetles are handsome insects and many species are adorned with bright iridescent colors. Close observation can often be challenging, however, due to their acute vision. If approached, they often fly several yards ahead in a kind of cat-and-mouse game. Tiger beetle larvae are also predatory. They construct burrows and wait patiently inside, popping out to ambush passing organisms. Once captured, the larva drags its prey into the burrow to feed.

Believe It or Not: Tiger beetles are the fastest terrestrial insects in the world, moving so quickly that their eyes cannot gather enough light to see. As a result, they have to pause periodically when running to locate their prey.

Beetles (Order Coleoptera)

Dung Beetle

Size: Highly variable; often 0.5–1 inch long
ID Tips: Compact and oval in shape; dull black to a variety of iridescent metallic colors; males often have horns
Range: Throughout the United States
Where to Look: On the ground

Yes, these interesting beetles love poop. The adults are highly mobile and fly to locate piles of fresh animal dung. Highly selective, many species prefer the dung of only certain animals, with many using that of herbivores, which primarily consists of undigested plant material. (You might also spot these little guys on your dog's droppings.) There are three basic types of dung beetles, each of which constructs a nest differently. These are rollers, tunnelers, and dwellers. The most unique are rollers. Resembling a miniature circus act, they form a bit of dung into a ball, physically roll it away, and eventually bury it in the soil. Tunnelers excavate beneath an existing dung pile, burying all or a portion of the feces. Lastly, dwellers simply live in the dung as-is, neither moving nor burying it. In all cases, the poop serves as food for their developing larvae. In the process of feeding, dung beetles serve to help clean up animal waste and recycle important nutrients. Adults may also be attracted to artificial lights.

Believe It or Not: Dung beetles locate and discriminate between poop types based on odor.

Beetles (Order Coleoptera)

On the Ground

Metallic Wood-boring Beetle

Size: 0.5–1.5 inches long

ID Tips: Streamlined, bullet-shaped body; a flat head; short antennae; a tapered rear; often have metallic iridescent colors

Range: Throughout the United States

Where to Look: On the ground or on trees, logs

Metallic wood-boring beetles are a highly attractive group named for the often-bright, lustrous iridescent colors seen in adults. This makes them prized by collectors. Active by day, they are often encountered at flowers, on vegetation, resting on logs, or even sitting on pathways. Adults are strong fliers and readily take to the air if disturbed. Female beetles seek out weak or damaged trees on which to lay eggs. They may attack a wide variety of host trees. The resulting larvae have powerful, well-developed jaws and bore under the bark and into the wood, often hastening the tree's demise. Infected trees have noticeable oval-shaped burrows that may be particularly obvious on fallen or harvested logs. Larval development can be slow, with many species requiring several years to mature.

Believe It or Not: Some wood-boring beetles can be extremely destructive pests. This is primarily true of invasive species such as the emerald ash borer, which was accidentally introduced into the United States from Asia in the 1990s.

Beetles (Order Coleoptera)

Bess Beetle

Size: 1.2–2.0 inches long
ID Tips: Robust glossy black body; short golden hairs on thorax and middle leg; wing cases appear deeply grooved
Range: The eastern United States
Where to Look: Under logs

These large insects are also called patent leather beetles due to their shiny black outward appearance. Only two species can be found in the United States, with both being restricted primarily to the eastern and central states. Bess beetles spend much of their lives in rotting logs. They use their powerful mandibles to tunnel through the wood to create the living chambers in which they will reproduce and rear their young. In the process, they serve as valuable decomposers in forest systems. Both adult male and female beetles provide parental care to the developing larvae, a very unusual habit for insects, let alone beetles. They feed the growing larvae macerated, pre-chewed wood and frass (poop) until fully grown. The resulting adult beetles will then leave the log, mate, and seek out new decaying logs to inhabit. They are often attracted to artificial lights at night.

Believe It or Not: Both adults and larvae are able to produce noticeable squeaking sounds when disturbed. They also use these sounds to communicate with one another, having specific sound signals for various purposes.

Beetles (Order Coleoptera)

On the Ground

Giant Stag Beetle

Size: 2.0–2.8 inches long

ID Tips: Large; shiny, dark brown to reddish-brown body; males with large pincher-like mandibles

Range: The eastern United States

Where to Look: On logs

Truly unmistakable in both size and appearance, the giant stag beetle is named for its long, bowed mandibles, which resemble antlers. Males use these enormous projections to joust with rival males for access to potential mates as well as to defend themselves from predators. Females are relatively similar in size but have much smaller jaws. Found in forests throughout the eastern United States, adults can often be found on dead logs or occasionally at artificial lights. Their larvae feed on decaying wood and develop in fallen trees or stumps, taking several years to reach maturity. **Be cautious:** If carelessly handled, adult beetles can deliver a strong pinch to fingers or hands.

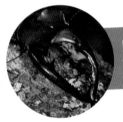

Believe It or Not: Despite their large mandibles, adult beetles have an entirely liquid diet and feed on tree sap and the juice from fermenting fruit.

Spiders (Order Araneae)

On the Ground

Carolina Wolf Spider

Size: Body 0.75–1.0 inch long; much longer legs
ID Tips: Hairy gray to dark brown body with darker markings
Range: Throughout the United States
Where to Look: On the ground

Resembling a small tarantula, the Carolina wolf spider is a somewhat menacing-looking creature. Adults can approach 4 inches across with their legs outstretched, making this species the largest wolf spider in North America. They are solitary, ground-dwelling organisms that are active nocturnal hunters. They rely on good vision and agility to run down or pounce on their prey instead of capturing organisms in a web. The Carolina wolf spider is common in a wide variety of habitats, including woodlands and grassy fields, but they may be easily overlooked because of their brown camouflaged appearance. They may occasionally wander into homes inadvertently, especially late in the year to seek shelter from cold temperatures. **Use caution:** While not aggressive and generally quite skittish of humans, they can inflict a painful bite if handled or provoked. Wolf spider eyes reflect light and can be easily spotted at night. See page 217 for details.

Believe It or Not: Female Carolina wolf spiders display unique maternal care, carrying their large, rounded egg sac on the back of their abdomens as they move about. When the young spiderlings hatch, they climb onto their mother's abdomen and ride along until they are old enough to venture off on their own.

Bees, Wasps, and Ants (Order Hymenoptera)

On the Ground

Pavement Ant

Size: 0.10–0.15 inch long
ID Tips: Small; body dark brown to black
Range: Primarily the northeastern United States
Where to Look: On the ground

Pavement ants are some of the most commonly encountered ants in the United States. Initially introduced from Europe more than a century ago, they have since spread to more than 25 states and are continuing to expand their range. They are most abundant in the Northwest, Midwest, and Northeast portions of the country. These tiny ants typically nest under stones, logs, or boards, but, as their name implies, are also frequently seen along cracks in driveways or sidewalks where they mine the loose soil to form small, rounded concave mounds around a central nest opening, which leads underground. They can form large colonies containing thousands of individuals. They feed on a variety of different foods, including dead insects, seeds, sugar, and fruit. They are able to forage some distance away from their nests and occasionally make their way into homes, often showing up in kitchens.

Believe It or Not: The pavement ant is one of the most abundant ant species in urban environments. Part of their success stems from their aggressive nature. They regularly battle other ants and overtake their territories, essentially laying claim to the highest-quality real estate.

Bees, Wasps, and Ants (Order Hymenoptera)

On the Ground

Red Imported Fire Ant

Size: 0.10–0.25 inch long
ID Tips: Small; body reddish-brown with a black rear end
Range: The southern United States
Where to Look: On the ground

This small but cantankerous insect is a severe nuisance and a major agricultural pest. Accidentally introduced from Brazil into the Deep South more than 70 years ago, the red imported fired ant has since spread across the southern United States from Maryland to southern California. Preferring dry, open areas, the ants construct conspicuous mounded nests along roadsides, in fields, pastures, lawns, and yards. The adults are extremely active and highly aggressive. They will quickly pour out of disturbed nests, attacking en masse to defend against any intruder. The ants can sting repeatedly, causing red, swollen blisters that both hurt and itch. **Be cautious:** Avoid stepping on or otherwise molesting nests. Imported red fire ants are omnivorous and feed on a wide variety of food, including other invertebrates, seeds, plants, dead animals, and honeydew. Their presence has been linked to severely reducing populations of ground-nesting birds, rodents, and reptiles.

Believe It or Not: In heavily infested locations, several hundred red imported fire ant colonies may occur per acre. Each colony in turn may have several hundred thousand occupants.

Bees, Wasps, and Ants (Order Hymenoptera)

On the Ground

Velvet Ant

Size: 0.25–0.8 inch long

ID Tips: Black or brown body marked primarily with red, yellow, or orange hair; an overall hairy appearance; resembles a large ant

Range: The southern United States

Where to Look: On the ground

Despite their name, velvet ants are actually brightly colored, solitary wasps. The winged males are active flower visitors and most often observed when feeding at available blossoms. Females lack wings and resemble large, furry ants. They are regularly spotted scurrying quickly along the ground in yards, pastures, or other open habitats. Velvet ants are parasitoids of various ground-dwelling insects. Females primarily seek out the subterranean nests of bees and wasps, laying their eggs on the developing brood. The hungry larvae start to feed and eventually consume their hosts before pupating. **Be cautious:** While not aggressive, female velvet ants can deliver a very painful sting if handled. Their vibrant colors signal danger and boldly advertise this formidable defense.

Believe It or Not: Many velvet ants make a noticeable squeaking noise when disturbed. This distinctive sound may help warn potential predators.

Bees, Wasps, and Ants (Order Hymenoptera)

Carpenter Ant

Size: 0.3–0.5 inch long
ID Tips: Black to reddish-brown with a rounded thorax; large jaws
Range: Throughout the United States
Where to Look: Under logs or on the ground

Several species of carpenter ants are found in North America. While they vary in size between species and even within individual colonies, they are some of the largest ants typically encountered. Living up to their name, carpenter ants nest in wood and use their large jaws to excavate elaborate tunnels. In the wild, they can be found in dead and decaying logs and trees, where they serve an important role in decomposition. However, they occasionally nest in homes or other structures, preferentially seeking out moist or water-damaged areas, where they can cause structural damage. Each colony has a queen and a large number of sterile workers, which perform various functions—from brood care and nest defense to general maintenance and food collection. Workers are most active outside the nest at night, when they are frequently encountered roaming along the ground in search of small insects.

Believe It or Not: Carpenter ants can produce large swarms of winged queens and males, all of which are capable of reproducing. This is particularly common in spring. After mating, each queen searches for a suitable site to start a new nest.

Butterflies and Moths (Order Lepidoptera)

On the Ground

Banded Woolly Bear Caterpillar

Size: 1.8–2.25 inches long

ID Tips: Densely fuzzy; banded with black at both ends and reddish-brown in the middle

Range: Throughout the United States

Where to Look: On the ground

The banded woolly bear may be the most beloved and well-known moth caterpillar in North America. It is distinctly furry-looking, with dense hairs over its body. The hairs on each end are black, while those in the middle are a rich reddish-brown, giving it an overall banded appearance. If disturbed, the larva will curl up in a tight ball and quietly wait for danger to pass. Banded wooly bear larvae are often spotted in late summer or fall when wandering quickly along the ground or over roads to find a protected location in which to overwinter. Once spring arrives, they will become active again, feed briefly, spin a cocoon, and soon emerge as a lovely Isabella tiger moth. While it is commonly believed that the width of the colored bands on larvae can predict the severity of the upcoming winter, their color variation is actually determined by many other factors, including age.

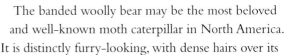

Believe It or Not: Banded woolly bear larvae produce increased quantities of glycerol in their blood prior to hibernation. This natural antifreeze helps them survive harsh winter temperatures.

Butterflies and Moths (Order Lepidoptera)

Mourning Cloak

Size: Wingspan 3.0–4.0 inches

ID Tips: Velvety black wings with a broad irregular yellow border and a row of bright purple-blue spots

Range: Throughout the United States and southern Canada

Where to Look: On the ground, on logs, or on branches

This lovely butterfly has a morbid name but is often the first harbinger of spring. Hibernating adults occasionally venture forth from log piles or other protected sites on warm winter days and fly about even with patches of snow still present. Adults are produced from a single generation emerge in early summer, lie dormant until fall, and become active again to feed and build up fat reserves before seeking sheltered locations to overwinter. Although common across the United States and southern Canada, they are typically encountered in very small numbers or as lone individuals. The adults have a strong, darting flight but frequently alight on fallen trees, tree trunks, or on the ground, where they tend to be quite wary and difficult to closely approach. The spiny gray and red larvae are gregarious and feed together on the leaves of a wide range of various trees, including willow, elm, birch, and hackberry.

Believe It or Not: Adults can survive for over 10 months, making the mourning cloak one of the longest-lived butterflies in North America.

151

Millipedes (Order Julida)

On the Ground

Millipede

Size: 0.5–6.0 inches long
ID Tips: Dark, smooth, cylindrical wormlike body with many small legs
Range: Throughout the United States
Where to Look: Under logs, rocks, or leaf litter

Millipedes are shiny, dark colored, wormlike creatures. Bearing two pairs of legs on each segment, they may have several hundred legs in total but nonetheless crawl slowly across the ground. Despite their somewhat creepy appearance, millipedes are completely harmless. They are, in fact, quite beneficial creatures, feeding on organic material and helping to recycle it. Found in a variety of dark, moist locations during the daytime, they are commonly encountered under logs, rocks, mulch, or even in flowerpots. They may from time to time inadvertently wander into homes or other structures, gaining access through small openings. When disturbed, millipedes tend to stop moving and tightly coil up in defense. It is common to also find dead and desiccated millipedes in this same posture on doorsteps, sidewalks, or in other open areas.

Believe It or Not: Millipedes may take several years to fully develop; the resulting adults live for several additional years.

Cockroaches and Termites (Order Blattodea)

American Cockroach

Size: 1.0–2.8 inches long
ID Tips: Oblong reddish-brown body; spiny legs and long antennae
Range: The southern United States and scattered other states
Where to Look: Under logs or leaf litter

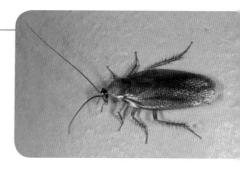

Cockroaches give insects a bad name. Often associated with waste, squalor, and disease, they are generally reviled and considered an unwanted organism that should be exterminated. Commonly found outdoors in moist, shady locations such as under logs, mulch, loose bark, or leaf litter, American cockroaches may occasionally make their way into homes in search of food or for protection from cold temperatures. They then usually become a nuisance. Adults and nymphs are opportunistic scavengers generally feeding on decaying organic material but may consume just about any available food items. While adults have wings, they seldom fly and most often scurry quickly along the ground or over objects, even occasionally climbing up into trees. Females lay eggs enclosed in a protective dark, oblong case and may produce dozens of egg cases over the course a lifetime.

Believe It or Not: Cockroaches have a pair of short appendages call cerci on the tip of the abdomen. These structures are extremely sensitive to the movement of air and enable the insect to quickly detect an approaching predator. This defensive strategy can make cockroaches quite challenging to capture.

Cockroaches and Termites (Order Blattodea)

On the Ground

Termite

Size: 0.2–0.6 inch long
ID Tips: Small whitish body with a somewhat darker head and short antennae
Range: Primarily the southern United States
Where to Look: Under logs

Termites are social insects and live in large colonies. They feed on plant material and dead wood with the help of symbiotic microbes in their guts, which enable them to break down cellulose into usable nutrients for food. In doing so, they play a highly beneficial role in the environment as decomposers. However, termites occasionally attack the wood in homes and other buildings, causing significant structural and economic damage. Nonetheless, termites though are most often encountered in dead trees, stumps, or decaying logs. Winged termites are frequently encountered, especially during swarming events. Some colonies can produce thousands of winged individuals. Swarming is part of the normal termite reproductive cycle and is typically triggered by colony age and suitable weather conditions. Winged adults congregate and mate with adults from other colonies. The resulting mated individuals fly away to start new colonies.

Believe It or Not: Unlike bees, wasps, and ants, termite colonies have both a king and a queen. The queen is responsible for egg laying and the king for mating with her at regular intervals to ensure the production of viable offspring.

Earthworms (Order Megadrilacea)

Earthworm

Size: 4.0–8.0 inches long
ID Tips: Elongated, cylindrical reddish-brown to pinkish body
Range: Throughout the United States
Where to Look: In soil; under logs; rocks or leaf litter

These organisms are long, segmented worms that are commonly found living in soil. Unlike insects, earthworms do not have an exoskeleton. They instead have a fluid-filled body cavity surrounded by muscle; this is known as a hydrostatic skeleton. A combination of fluid pressure and muscle action keeps the body's shape and enables it to move. As they breathe through their skin, earthworms require moist environments and thus may also be frequently encountered under logs, rocks, or leaf litter. More than just fish bait, they are highly beneficial decomposers, feeding on—and breaking down—organic matter, which greatly enhances soil fertility. Just the mere underground movement of earthworms also helps improve soil structure and aeration, enabling water, air, and nutrients to enter. Earthworms have both male and female sex organs but need to mate to reproduce.

Believe It or Not: Earthworms are often spotted in good numbers on rainy days. Despite the common belief that they are flooded out of their underground burrows, scientists believe that his behavior actually represents migration. The rain enables the worms to move farther than they otherwise could.

Vinegaroons (Order Uropygi)

Vinegaroon

Size: 1.5–2.5 inches long

ID Tips: Body dark brown to black; six walking legs; a pair of long modified front legs that resemble antennae; two enlarged front pinchers; a long, thin tail

Range: Primarily found in the Southwest; also found in Florida

Where to Look: Under logs or rocks

The vinegaroon is an alien-like creature that looks repulsive when first encountered. Limited in range, it's found in the desert Southwest from Arizona to Texas and Oklahoma. It's also found in Florida. Also called a whip scorpion, it has a long, slender tail that can be used to spray concentrated acetic acid—vinegar—in defense. The resulting vinegar smell gives the vinegaroon its distinctive name. A nocturnal predator, adults have poor vision but use their modified pair of front legs as sensory organs to help locate available arthropod prey. Once in range, they use their large and powerful pedipalps (pinchers) to capture and crush unfortunate organisms before consuming them. During the daytime, vinegaroons reside in underground burrows or under logs, rocks, or boards. If disturbed, they often rear up and spread their pedipalps in a defensive posture.
Be cautious: While unable to sting or bite, they can inflict a painful pinch.

Believe It or Not: Slow growing and fairly long-lived, vinegaroons can live for up to seven years.

Lacewings, Mantidflies, Antlions (Order Neuroptera)

Ant Lion Larva

Size: 0.35–0.45 inches

ID Tips: Small pits in sandy soil that resemble inverted cones; one larva is located at the base of each pit

Range: Throughout the United States

Where to Look: In the ground

Also called doodlebugs, antlion larvae are voracious predators. Brown in color, they have an oval to rounded body with large, ferocious-looking jaws, which they use to grab prey. Individual larvae construct distinctive funnel-shaped pits in open dry, loose soil or sand, each being approximately one to two inches in diameter. If the conditions and terrain are right, there can be quite a few in one area. Once complete, they patiently wait, concealed by soil, at the bottom of each pit for unsuspecting ants or other small insects to inadvertently tumble to the bottom. The ant lion larva then quickly grabs the organism, paralyzes it with venom, and sucks out the liquefied body contents. The desiccated carcass is then flipped out of the pit and the larva resumes its wait for another victim.

Believe It or Not: Ant lion larvae are easy and highly entertaining insects to keep in captivity. Scoop the pit out until the larva is located. Place the larva in a small, partially sand-filled bowl or cup at least 4-6 inches across. Once its pit is constructed, simply drop in an ant, spider, or other small insect every few days. No water is needed.

Snails (Order Pulmonata)

On the Ground

Snail

Size: Highly variable; 0.2–8.0 inches long
ID Tips: Brown to dull-colored fleshy body with elongated stalklike tentacles stemming from the head; a hard shell on the back
Range: Throughout the United States
Where to Look: Under logs or leaf litter

Snails are odd, somewhat slimy invertebrates that are often quite common in gardens and yards. Like slugs, they have soft, unsegmented bodies. Snails also sport a hard shell for protection from both predators and harsh environmental conditions. As a snail grows in size, it continues to expand its shell until fully grown. These terrestrial mollusks can be found in a wide variety of locations from forests to cities. They tend to prefer moist conditions with an abundance of organic material, living among leaf litter, plants, fungi, and dead wood. Through the process of feeding, they break down these resources, helping produce nutrient-rich soil. Snails tend to be most active at night and often seek shelter during the daylight hours in cool, dark locations. When disturbed, they rapidly retract their soft bodies into the protective hard shell. Snails also leave behind a characteristic trail of slime; this slime serves as a lubricant and prevents damage to their fragile body.

Believe it or not. Snails actually move faster than one might think, approximately one meter per hour. They also save energy by reusing existing slime trails laid down by other snails.

Slugs (Order Soleolifera)

On the Ground

Slug

Size: 1.0–4.0 inches long

ID Tips: Brown to dull colored; fleshy body is often patterned; elongated stalklike tentacles protruding from the head; lacks hard shell

Range: Throughout the United States

Where to Look: On the ground

These fleshy, somewhat nondescript critters are terrestrial mollusks. Like snails, slugs have unsegmented bodies and secrete mucus or "slime" to help prevent them from drying out and to facilitate movement. They also feed on a wide variety of living and dead organic material, including plants, fruit, and fungi. Unlike snails, slugs lack an external hard spiral shell. They tend to be regularly encountered in gardens, among leaf litter, and other often-moist locations with available food resources. They can occasionally be significant plant pests. They are most active at night or on cool, rainy days. Both snails and slugs are hermaphrodites, having both male and female reproductive organs. This means that all slugs can produce fertile eggs after mating.

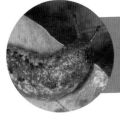

Believe It or Not: Slugs spend much of their lives underground in moist soil; there they are protected from dehydration and the many potential predators aboveground.

Earwigs (Order Dermaptera)

On the Ground

Earwig

Size: 0.5–0.75 inch long

ID Tips: Elongated reddish-brown body; prominent pinchers on the end of the abdomen

Range: Throughout the United States

Where to Look: Under logs, rocks, or leaf litter

Despite the widespread lore, these ancient-looking insects do not burrow into the ears of unsuspecting people. Earwigs are common nocturnal creatures that scavenge for food or prey upon other small organisms. They also occasionally feed on plants. During the day, they typically hide under logs, amid leaf litter, or in other moist, dark locations. They regularly wander into homes or garages. Earwigs have distinctive pincher-like features (known as cerci) on their rear end. These structures are more pronounced and curved in males. The cerci are used for defense, as well as to help capture available prey. There are more than two dozen earwig species in the United States, several of which are non-native.

Believe It or Not: Female earwigs are good mothers. They actively protect their nest and provide food for their developing young.

Harvestmen and Daddy Longlegs (Order Opiliones)

Harvestmen and Daddy Longlegs

Size: 0.15–0.3 inch long; much longer legs
ID Tips: Small, round brownish body; eight extremely long thin legs
Range: Throughout the United States
Where to Look: On the ground or on trees or logs

These fragile-looking organisms are also commonly called daddy longlegs. Despite their outward appearance, they are not actually true spiders but instead are more closely related to scorpions. Harvestmen have a round or oval compact body that lacks any noticeable division, quite unlike the typical two-part spider body plan (which consists of a cephalothorax and a distinct abdomen). Additionally, most spiders possess eight eyes; harvestmen have only two. Also, unlike spiders, harvestmen do not have fangs or produce silk. They can be found in a wide range of different habitats but often prefer moist locations under objects or on shady leaves. Harvestmen are generally nocturnal scavengers or predators and feed on a broad array of different items, from insects, worms, and slugs to fungi and decaying plant material. Periodically, they may cluster together in large interwoven masses in protected locations on trees or other objects. This behavior is most common in the fall and is the primary reason for these creatures' common name of harvestmen.

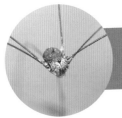

Believe It or Not: The legs of harvestmen may break off easily when handled and are often sacrificed in order to escape a predator.

Centipedes (Class Chilopoda)

Centipede

Size: 1.0–5.0 inches long
ID Tips: Typically tan to dark brown; segmented wormlike bodies with many long legs
Range: Throughout the United States
Where to Look: Under logs, rocks, or leaf litter

Centipedes are elongated, flattened organisms with a truly alien-like appearance. Although their name literally means 100 legs, most species have significantly fewer. While superficially similar to millipedes, centipedes have only one pair of legs per segment, which extend outward from their body. Requiring moist locations, they are often found under logs, rocks, old boards, or leaf litter but may occasionally be seen in damp basements, garages, or crawl spaces. When disturbed, centipedes swiftly scurry along the ground in a somewhat snakelike sinuous motion. They are ferocious nocturnal predators and use their speed and venomous claws to quickly subdue available prey, feeding on an assortment of small insects, worms, or other available critters. **Take caution:** Although bites are uncommon, they can be quite painful, so care should be taken to avoid handling individuals.

Believe It or Not: While the majority of centipedes are small, some species found in the desert Southwest are true giants and may approach eight inches in length.

Grasshoppers, Crickets, and Katydids (Order Orthoptera)

Field Cricket

Size: 0.75–1.2 inches long
ID Tips: Dark brown to black with membranous wings; enlarged hind legs; long antennae; two prominent tail filaments
Range: Throughout the United States
Where to Look: On the ground

These musical insects are more often heard than seen. Uniformly dark brown to black, field crickets can be found in open fields, lawns, forests, and occasionally even in houses or basements. They have large hind legs for jumping and can quickly escape if approached. Adults and nymphs are omnivorous and quite opportunistic diners. They feed on living plants, dead organic material, fruit, and other insects. Only male crickets sing, producing distinct pulses of noise either as short chirps or as longer trills, depending on the species. They produce their calls by rubbing their wings together in rapid succession, and their song is a courtship melody used to attract females. After mating in the fall, most females lay hundreds of eggs in the soil using a long, pointed ovipositor. The eggs overwinter and hatch the following spring.

Believe It or Not: Parasitic flies have learned to eavesdrop on male courtship calls in order to identify the location of a cricket host; they then lay their eggs on it, and the resulting fly larvae burrow into the cricket's body, feed on its internal tissues, eventually killing it.

Aphids, Cicadas, and Others (Order Hemiptera)

On Vegetation

Woolly Aphid

Size: 0.08–0.15 inches
ID Tips: Small; pear-shaped body covered with white waxy filaments
Range: Throughout the United States

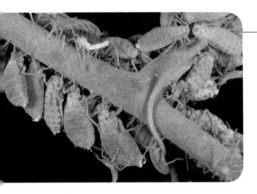

These bizarre creatures appear to be perpetually having a bad hair day. Like other aphids, they feed in colonies on plant sap and generate a sugary waste product called honeydew. Woolly aphids are most famous for producing the waxy secretions that cover their bodies and give them an overall fuzzy, cottonlike appearance. These filaments help protect these vulnerable insects from predators. Woolly aphids generally feed on two different plant species throughout a given year. They overwinter on one host and begin feeding on its leaves or stems during the spring months. The colony grows as females produce offspring by live birth, each resulting insect being exact clones of its mother. Soon winged individuals develop and move to a secondary host plant, start feeding, and produce more offspring over the summer months. As fall nears, they return to the primary host, mate, and lay eggs on the plant. The cycle begins again the following spring.

Believe It or Not: Woolly aphids are often preyed upon by harvester butterfly larvae, the only carnivorous butterfly species in North America.

Aphids, Cicadas, and Others (Order Hemiptera)

On Vegetation

Aphid

Size: 0.10–0.15 inch long

ID Tips: Very small, oval body with a pair of pipelike projections on the abdomen; color varies from green and yellow to orange or even black

Range: Throughout the United States

These common pear-shaped insects are the scourges of most gardeners. Easily overlooked due to their minute size, aphids often look like small spots or eggs on vegetation. Upon closer inspection, often with a magnifying glass, their true identity is revealed. They typically occur in colonies and use their piercing-sucking mouthparts to feed on plant juices, often preferring new leaves or buds. Their colonies grow quickly, and heavy infestations can cause damage to plants, mostly in the form of discolored or malformed leaves. Some aphids even transmit diseases to plants. In the process of feeding, aphids secrete a sugary liquid byproduct called honeydew that is fed upon by other insects, including wasps, butterflies, and ants. In return for this sugar-rich treat, ants often protect and care for their aphid companions, creating a mutually beneficial relationship.

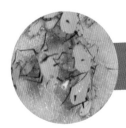

Believe It or Not: Female aphids are capable of producing live young instead of eggs.

Aphids, Cicadas, and Others (Order Hemiptera)

On Vegetation

Planthopper

Size: 0.10–0.25 inch long

ID Tips: Variable in color; white, brown, or green; broad wings held vertically; some distinctly wedge-shaped

Range: Throughout the United States

Like aphids, planthoppers are small insects that feed on plant juices and excrete honeydew as a waste product. While their feeding typically does little mechanical harm to plants, it can transmit disease and can cause serious damage to some economically important crops. Planthoppers are a diverse group, with more than 900 species found in the United States. Many adults tend to have broad, transparent wings that often resemble small leaves. They tend to rely on camouflage for protection, often remaining still or moving slow to avoid detection but can readily hop if needed, as their name suggests. Most nymphs are gregarious, as are some adults, and they often form large clusters on plants.

Believe It or Not: The nymphs of some planthoppers produce waxy secretions from glands in the abdomen, forming elaborate filaments that project off the back of the body or cover it in a fluffy costume-like coating for disguise.

Aphids, Cicadas, and Others (Order Hemiptera)

On Vegetation

Leafhopper

Size: 0.15–0.75 inch long

ID Tips: Narrow wedge-shaped bodies; color varies tremendously; some are green to brown, whereas others are brightly colored

Range: Throughout the United States

Leafhoppers are a diverse group of primarily small, plant-dwelling insects. Both adults and nymphs feed on plant sap, using their piercing-sucking mouthparts to penetrate the living tissues of leaves and stems. While their feeding generally causes little damage, there are several species that are considered serious crop pests. Leafhoppers are common in gardens and suburban landscapes where their small size and often-camouflaged appearance makes them easy to overlook. They have excellent vision and readily scurry away when disturbed, crawling quickly along vegetation or jumping with powerful hind legs, as their name suggests. The adults of many species can also fly. As a result, leafhoppers can often disperse long distances to colonize new locations.

Believe It or Not: Leafhoppers communicate with each other using vibrations generated from their abdomen. These signals, too faint for humans to hear, travel along the surface of vegetation but may also be audible to leafhoppers on nearby plants.

Aphids, Cicadas, and Others (Order Hemiptera)

On Vegetation

Tarnished Plant Bug

Size: 0.18–0.25 inch long

ID Tips: Oval, somewhat flattened yellow-brown body with dark markings; long antennae

Range: Throughout the United States

These small insects are members of a large and diverse family of true bugs that primarily feed on plants. The tarnished plant bug gets its name for its somewhat dull appearance, which is similar to tarnished metal. These bugs are widespread and abundant and boast an extensive host range, able to utilize hundreds of different plants, including many crop species. Because they prefer feeding on new growth, flowers, and developing fruit, adults and immature specimens can cause significant damage, especially when present in high numbers. They can also transmit various plant diseases. Highly mobile insects and good colonizers of new host resources, they produce multiple generations each year. As winter approaches, the adult bugs seek shelter in plant debris, leaf litter, or other protected sites, hibernating until the following spring.

Believe It or Not: The tarnished plant bug is considered a serious pest and feeds on some 130 different economically important plant species.

Aphids, Cicadas, and Others (Order Hemiptera)

Spittlebug

Size: 0.25–0.4 inch long

ID Tips: White frothy mass on plants; the insect feeds inside, hidden from sight

Range: Throughout the United States

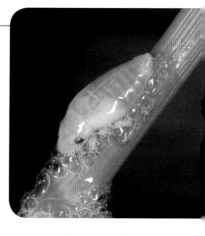

Spittlebugs might be the most common insects that most people haven't actually seen. Nonetheless, most of us however are familiar with the white frothy foam sometimes seen on vegetation. Inside this bubbly mass is an immature spittlebug. The developing pale nymphs feed on plant juices and pump air into their fluid secretions, creating the spittle-like mass that serves both to protect them and prevent dehydration. The resulting adults, called froghoppers, are good jumpers and can also fly. They are often attracted to artificial lights at night. One or more generations are produced each year, depending on latitude, with the eggs overwintering until the following spring.

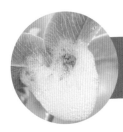

Believe It or Not: The spittle surrounding the nymph also acts like insulation, helping buffer the insect from temperature extremes.

Aphids, Cicadas, and Others (Order Hemiptera)

On Vegetation

Treehopper

Size: 0.25-0.5 inch long

ID Tips: Highly variable, from greens and browns to multicolored; typically resemble small leaves or thorns

Range: Throughout the United States and Canada

Treehoppers take the art of camouflage to the next level, occurring in a tremendous variety of forms and colors. They have an enlarged pronotum—a platelike structure that covers their thorax. Looking something like an ornate bicycle helmet, it helps them blend in with vegetation and enables them to resemble thorns, leaves, buds, or other plant projections with amazing precision. With more than 250 species found in the United States and Canada, no single description adequately captures this fascinating insect group. Treehoppers feed on plant juices and are often generalists, utilizing a wide range of different hosts. Many species are gregarious and often form large, and somewhat prominent, groups. Typically one or more generations are produced each year. Female treehoppers cut small slits in branches and insert their eggs within. The eggs safely overwinter inside and hatch the following spring.

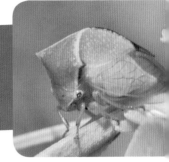

Believe It or Not: Many treehoppers communicate with each other using low-frequency vibrations generated from their abdomens. These vibrations resonate along plant stems and are used to attract and court mates, signal to rival males, or alert others that a predator is near.

Aphids, Cicadas, and Others (Order Hemiptera)

Boxelder Bug

Size: 0.4–0.5 inch long
ID Tips: Dull black flattened oval body; reddish-orange stripes on the thorax and wing bases
Range: Throughout the United States

As its name suggests, this colorful bug is often closely associated with boxelder trees, its primary host. Both adults and nymphs have piercing-sucking mouthparts and feed on the sap from the leaves and developing seeds. Late in the season, they seek out protected areas in which to overwinter; they may often congregate on the sunlit sides of homes and buildings in large numbers. They then often make their way into walls, attics, or other hollow spaces via small cracks or gaps; there they hibernate and are seldom seen. However, on mild days, boxelder bugs may become active and venture inside, often appearing on sunny windowsills or baseboards in some numbers. While they pose no direct harm to humans or structures, their unexpected presence is often alarming and bothersome. In spring, they leave their protected sites and make their way back to host trees to reproduce.

Believe It or Not: Boxelder bugs release a pungent odor when disturbed or crushed. The bright colors of adults and nymphs signal to predators that they are chemically protected.

Aphids, Cicadas, and Others (Order Hemiptera)

On Vegetation

Large Milkweed Bug

Size: 0.4–0.6 inch long
ID Tips: Elongated orange-and-black body; found on milkweed plants
Range: Throughout the United States

As its name implies, the large milkweed bug can readily be found feeding on milkweed plants. It uses its piercing-sucking mouthparts to sip juices from the leaves, stems, and developing seedpods of its host. In so doing, it absorbs toxic chemicals from the plant into its body, which helps prevent it from being eaten by predators. And just like its more famous insect relative, the Monarch butterfly, the large milkweed bug advertises this defense by its bold orange-and-black warning coloration, which signals danger to potential predators. Large milkweed bugs often feed in small groups, especially when young, and this helps advertise the display. They are wary of close approach and readily scurry or even drop off the plant if disturbed.

Believe It or Not: Large milkweed bugs are easy to rear in captivity. They are available commercially for home or school use and readily feed on sunflower seeds.

Aphids, Cicadas, and Others (Order Hemiptera)

Spined Soldier Bug

Size: 0.45–0.6 inch long

ID Tips: Mottled brown color; a shield-shaped body, one spine on each shoulder and a dark diamond on the hind end where the wings overlap

Range: Throughout the United States

Named for its prominent, pointed shoulders and ravenous predatory behavior, this drab insect is one of the most common beneficial stinkbugs in North America. Found coast to coast, it is a generalist hunter in all life stages and scours vegetation in search of available prey, feeding on an extensive range of insects, including many undesirable pest species. They are highly mobile creatures and readily move from plant to plant. The spined soldier bug uses its sharp piercing-sucking mouthparts to harpoon its victims, inject digestive enzymes, and slurp up the liquefied body contents. Both adults and their nymphs are capable of taking on organisms much larger than themselves. As a result, they are regularly sold by many retailers to help provide natural pest control for home gardens and commercial facilities alike.

Believe It or Not: Spined soldier bug nymphs are gregarious, especially when young. If insect prey is scarce, they have a tendency to feed on one another.

Aphids, Cicadas, and Others (Order Hemiptera)

On Vegetation

Green Stink Bug

Size: 0.5–0.7 inch long
ID Tips: Bright green body that is flattened and shield-shaped
Range: Throughout the United States and southern Canada

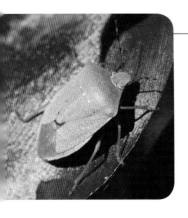

This common insect is found throughout the United States and southern Canada. The bright-green adults are readily encountered on vegetation, where they use their beaklike piercing-sucking mouthparts to pierce plant tissues and suck out the internal fluids. As a result, they can cause plant injury when numerous and are considered a crop and garden pest, feeding on a wide range of hosts from ornamental trees to vegetables. The adults have wings and are highly mobile, moving from one plant to another with a somewhat clunky, droning flight. Females lay clusters of compact, barrel-shaped eggs on vegetation. The resulting nymphs often feed together on plant material. They are oval-shaped, wingless, and have a mottled red, green, and black color pattern. As fall approaches, the green stinkbug seeks sheltered locations in which to overwinter, emerging in spring to start the cycle over again.

Believe It or Not: Stinkbugs get their unflattering name from their defensive behavior of releasing a foul-smelling chemical from specialized abdominal glands.

Aphids, Cicadas, and Others (Order Hemiptera)

Leaf-footed Bug

Size: 0.75–1.0 inch long
ID Tips: Narrow brown body; long antennae and prominent leaflike projections on their hind legs
Range: Throughout the United States; most species in the South

Leaf-footed bugs belong to a large and diverse insect family that includes many pest species. They tend to be common and often quite conspicuous due to their size. They have long piercing-sucking mouthparts, which they use to pierce leaves, stems, seeds, and fruit and drink the plant juices. They get their unique name from the distinctive flattened projections on their hind legs; these features resemble leaves and help them blend in with plants. Adults are readily encountered on vegetation. While they generally scurry when approached, adults are strong fliers and often make a dull droning noise when in the air. Several generations are produced each year. Adult leaf-footed bugs that emerge in late summer overwinter, often in large groups.

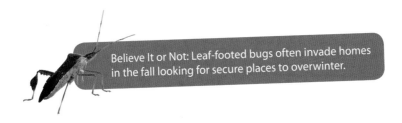

Believe It or Not: Leaf-footed bugs often invade homes in the fall looking for secure places to overwinter.

Aphids, Cicadas, and Others (Order Hemiptera)

On Vegetation

Cicada Nymph Exoskeleton

Size: 0.8–1.1 inches long
ID Tips: Translucent brown stout body with an opening down the back
Range: Throughout the United States

Cicadas are robust insects that look somewhat like giant flies. They are most often heard and not seen, spending much of their time high in the branches of trees and shrubs. However, most people have encountered old exoskeletons that have been shed by nymphs. Cicada nymphs develop underground, where they use their piercing-sucking mouthparts to pierce the roots of various trees and shrubs and feed on the plant juices. Depending on the species, it can take them between 2 and 17 years to reach maturity. When ready, the nymphs emerge from the soil and climb up onto trees or even the sides of homes or buildings. They then molt their skin one last time and become winged adults. Like a snapshot in time, the old, hollow brown papery exoskeleton is left behind along with a noticeable hole in the back, which is where the adult emerged.

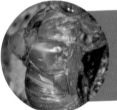

Believe It or Not: Certain cicada species, such as the famous periodic cicadas that emerge at a given number of years, may emerge in tremendous numbers. When they do, you'll notice: the adults and the old shed exoskeletons of the nymphs can literally cover tree trunks and branches.

Aphids, Cicadas, and Others (Order Hemiptera)

Wheel Bug

Size: 1.0–1.25 inches long
ID Tips: Gray-brown body with long legs and antennae; a narrow head; a prominent spiny rounded ridge on the thorax
Range: Throughout most of the United States

This bizarre and beneficial insect is the largest assassin bug found in North America. Distinctive in appearance, it has a prominent rounded crest that resembles a mohawk haircut and can be readily identified by this single feature. Like other assassin bugs, it uses its long and sharp beaklike mouthparts to quickly stab nearby unsuspecting prey. The wheel bug's saliva contains a toxin that paralyzes the unfortunate victim within seconds; the wheel bug then sucks out the liquefied body contents. Both adults and immature specimens are voracious hunters and readily attack a wide range of insects, including many potential garden pests, such as aphids, grasshoppers, other true bugs, as well as caterpillars.
Be careful: While not aggressive, wheel bugs can inflict an excruciatingly painful bite if handled.

Believe It or Not: Wheel bugs can defend themselves by turning a pair of bright orange, fleshy scent organs from their rear end inside out, which releases a foul, pungent odor.

On Vegetation

Aphids, Cicadas, and Others (Order Hemiptera)

Cicada

Size: 1.0–2.0 inches long

ID Tips: Stout, robust body; conspicuous, widely separated eyes; clear transparent wings

Range: Throughout the United States

Cicadas are large, stout insects that somewhat resemble giant flies. Adults spend much of their time out of sight in trees, but they produce loud, almost deafening courtship calls that can be heard on summer days. Immature cicadas, called nymphs, live underground where they feed on the sap from plant roots. When development is almost complete, the nymph emerges from the soil, climbs up a nearby tree or structure, and molts for the last time, becoming an adult. The papery brown shell of the old nymph remains behind. While more than 100 species are found in North America, the most unique species are famous for their 13- or 17-year life cycles. In the years when they emerge, they appear in huge numbers, sometimes exceeding 1 million cicadas per acre. This short-lived bounty provides a feast for birds, small mammals, and other wildlife.

Believe It or Not: Cicadas are among some of the loudest insects in the world, often exceeding the noise produced by a jackhammer or a power mower.

Beetles (Order Coleoptera)

On Vegetation

Weevil

Size: 0.10–0.25 inch long

ID Tips: Typically dark, oval to pear-shaped body; distinctive snout off the head and elbowed antennae

Range: Throughout the United States

Weevils are a diverse group of relatively small beetles, with more than 60,000 species worldwide. They are easy to identify because of their elongated head and conspicuous snout, which has a chewing mandible at the end. The length of the snout is highly variable. Some species have mere stubs, while others possess exaggerated snouts that may exceed the length of the body. Weevils are herbivorous and feed on both living and decaying plant material. They use their snouts for feeding and to create cavities in which to lay eggs. Many species are extremely destructive pests and cause millions of dollars in damage; they can affect numerous crops, including various fruits, berries, nuts, vegetables, and grains. The boll weevil is one of the best-known examples and is responsible for reshaping agriculture in the Deep South.

Believe It or Not: Weevils play dead when disturbed. They tuck their legs in, fall off a plant, and remain motionless for some time before venturing back to the vegetation.

Beetles (Order Coleoptera)

On Vegetation

Tortoise Beetle

Size: 0.20–0.25 inch long

ID Tips: Small and round to oval flattened body; typically metallic in color with a shell-like covering over the head and legs

Range: Throughout the United States

As their name suggests, these distinctive tiny beetles have body margins that extend over their head and legs, closely resembling the shell of a turtle. They are plant feeders, nibbling away at the leaves of various plants. Although tortoise beetles are often quite colorful and metallic in appearance, their small size makes them easy to overlook. The distinctive spiny larvae attach debris, old shed skins, and even fecal material to their backs as shields and to protect themselves from predators. Several generations may be produced each year. As cool weather approaches, adult beetles seek out dry, protected places, burrowing into leaf litter or beneath tree bark to spend the winter.

Believe It or Not: Some tortoise beetles have the amazing ability to change color. They do so by altering the fluid in their exoskeleton; this influences its overall reflectivity and the metallic-like appearance.

Beetles (Order Coleoptera)

Lady Beetle

Size: 0.25–0.4 inch long
ID Tips: Oval; orange to red with black spots
Range: Throughout the United States

Also called ladybugs or ladybird beetles, these small, shiny beetles are some of the most familiar and commonly encountered insects. They are a diverse group, with more than 400 species found in the United States. While appearance varies between species, most have some combination of the recognizable orange, red, or black-spotted color pattern. Generally, lady beetles are highly beneficial as both adults and larvae. They are voracious predators of a wide range of small insects, including many plant pests, including whiteflies, scale insects, and mealybugs, but they are probably best known for their appetite for aphids. The elongated larvae have been said to resemble tiny alligators and are adept hunters. Lady beetles typically have several generations each year. As fall approaches, many adult beetles seek sheltered sites in which to overwinter.

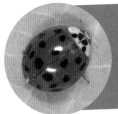

Believe It or Not: Adults of the introduced multicolored Asian lady beetle often invade homes or other structures in the winter by the thousands. They become active again in spring or on warm winter days and are often spotted in windows, along baseboards, or on walls or ceilings. They also have the nasty habit of delivering a somewhat painful bite.

Beetles (Order Coleoptera)

On Vegetation

Japanese Beetle

Size: 0.3–0.5 inch long

ID Tips: Metallic green with copper-colored wing coverings and white tufts along the abdomen; often feeds in groups

Range: Mostly in the eastern United States; sporadically westward

As its name suggests, this shiny, oval-shaped beetle is native to Japan. It was accidentally introduced into the United States in 1916 and has since spread to some 30 states. Like other invasive species, it has taken full advantage of its new home, which is packed with a bounty of food resources and a lack of natural enemies, enabling it to become a serious plant pest and cause millions of dollars in damage each year. While the adults chew away at the leaves, flowers, and fruit of various crop and ornamental species, their white larvae or grubs live underground, consuming the roots of lawn grasses. In most areas, Japanese beetles complete their life cycle in a single year. Adults emerge in summer and feed and reproduce before dying off as cold weather approaches.

Believe It or Not: Japanese beetles can feed on more than 300 different plants.

Beetles (Order Coleoptera)

Red Milkweed Beetle

Size: 0.35–0.6 inch long
ID Tips: Red body with black spots, black legs, and long black antennae
Range: The eastern United States

These bright-colored insects are always found in association with milkweed plants. In fact, the beetle's entire eastern range directly overlaps with that of the common milkweed, its dominant host. The adult beetles are herbivores and feed on milkweed leaves and developing flower buds. Like the monarch butterfly, the red milkweed beetle absorbs toxic chemicals into its body through the feeding process, rendering them distasteful to many predators. Its distinctive, bold red-and-black pattern clearly showcases this potent chemical defense. Like other long-horned beetles, it boasts a pair of long, black antennae. Female beetles lay their eggs on, or in, milkweed stems. The resulting larvae feed within roots where they overwinter, completing development the following spring.

Believe It or Not: When handled, adult beetles often make a noticeable squeaking noise.

Beetles (Order Coleoptera)

On Vegetation

Net-winged Beetle

Size: 0.45–0.6 inch long

ID Tips: Variable color; typically orange and black; soft, somewhat teardrop-shaped, ridged wing cases and long, segmented antennae

Range: Throughout the United States

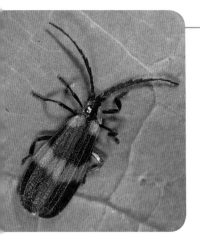

These colorful insects can be identified by the highly ridged and soft, often baggy appearance of their wing cases. Net-winged beetles tend to be chemically protected from predators and boldly advertise their distasteful nature with bright black, red, and orange color patterns. Typically forest dwellers, look for them resting on vegetation along shady trails. Their larvae also display bright contrasting colors and have been known to assemble in large aggregations to pupate. It is believed that these mass groupings help let predators know they are toxic and to be avoided. Overall, much of their biology and behavior is poorly known.

Believe It or Not: Many other insects, including other beetles, moths, and even some flies, mimic net-winged beetles for protection from predators.

Beetles (Order Coleoptera)

On Vegetation

Green June Beetle

Size: 0.7–1.0 inch long
ID Tips: Stout, oval, metallic green body with rust-colored stripes
Range: The eastern United States

This is a large, robust beetle common throughout much of the eastern United States. Considered a destructive pest and a nuisance, the adult beetles feed on various fruits, including many crop species, preferring those that are mature or overripe. They are active fliers and commonly encountered in open, grassy locations where they mate and lay eggs. Females burrow into the soil to deposit the eggs. The resulting white larvae or grubs tunnel in the ground and feed on decaying organic material, often near the soil surface. Extensive tunneling can disrupt the contact between plant roots and the soil, causing increased plant stress and even death. The nearly mature grubs overwinter deep in the soil and pupate the following spring. Adult beetles begin to emerge in June, thereby earning the common name of the green June beetle.

Believe It or Not: Green June beetle larvae are nocturnal and have the unusual habit of coming out of the soil at night and moving from one location to another; they move by crawling along on their backs.

Butterflies and Moths (Order Lepidoptera)

On Vegetation

Inchworm

Size: Highly variable; 0.5–1.5 inches long

ID Tips: Generally green to brown; smooth, elongated body with legs at both ends but none in the middle

Range: Throughout the United States

Inchworms are the larvae of geometer moths, a large and very diverse family with more than 1,400 species occurring in North America. They are distinctive in appearance; they have short legs at both ends of their elongated body but none in the middle. As the endearing childhood song mentions, an inchworm moves by extending its front end forward, grasping a leaf or twig, and then looping its body to pull along its rear end, almost as if it's literally measuring the distance. They tend to be drab colored and highly camouflaged, as this helps them blend in with the surrounding vegetation as they feed. They may often resemble a dead twig or leaf stem. While geometer moths are highly variable in color and pattern, most are quite small in size and characteristically rest with their wings spread. They are common around artificial lights at night.

Believe It or Not: Inchworms utilize a unique tactic to avoid being eaten. If they detect vibrations from an approaching predator, larvae often anchor a silken lifeline to nearby vegetation and jump off the plant on which they were feeding. As a result, they may frequently be spotted dangling in the breeze from overhanging branches.

Butterflies and Moths (Order Lepidoptera)

Tussock Moth Larva

Size: 1.0–1.5 inches long

ID Tips: Hairy, gray-to-cream body with two long black hairs extending from past the head and one from the rear end; four compact bushy hair tufts on the back

Range: The eastern United States

While most people seldom pay much attention to adult tussock moths, their fluffy larvae are a common sight in most yards. They are named for the four distinctive thick tufts of hair on their back; these resemble grass clumps, which are also known as "tussocks." The developing caterpillars feed on a variety of trees and can be pests if numerous, sometimes causing extensive defoliation. Full-grown larvae often crawl down the tree and wander extensively while looking for a suitable place to spin their cocoons. During this time, they are often seen on sidewalks, driveways, or on the sides of houses. **Be cautious:** If handled, the hairs on the larvae can cause skin irritation and itching. The cocoons may also cause the same reaction.

Believe It or Not: Female tussock moths can't fly. They remain on their cocoon to mate and lay eggs. Male moths have fully developed wings and are readily attracted to artificial lights.

Butterflies and Moths (Order Lepidoptera)

On Vegetation

Bagworm

Size: 1.0–2.0 inches long
ID Tips: Brown, cocoon-like structure found hanging from a branch
Range: Throughout the United States

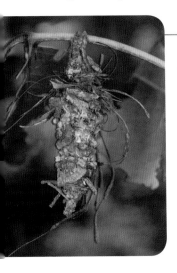

Bagworms are moth caterpillars, but they act like miniature hermit crabs because they carry around a protective home on their backs. They construct this "bag" by weaving twigs, leaves, or other debris together with silk. The bag is expanded as the caterpillar grows. Disguised in its well-camouflaged structure, the bagworm is free to move about on the plant and feed—and hopefully avoid the keen eye of a hungry predator. When fully grown, the caterpillar firmly attaches the bag to a branch with silk, pupating inside. The drab male moths emerge in the fall and look for mates. While males have fully developed wings, all females are flightless and remain in their bags, even laying all their eggs inside to overwinter. Only one generation is produced each year. Despite their novelty, bagworms can occasionally defoliate or even kill their host plants during severe outbreaks.

Believe It or Not: Newly hatched caterpillars often disperse to other plants by spinning strands of silk. Acting like a balloon in the breeze, these tiny threads catch the wind, carrying the larvae to a new location.

Butterflies and Moths (Order Lepidoptera)

On Vegetation

Fall Webworm

Size: Larva 1.25–1.5 inches

ID Tips: Geographically variable; hairy yellow-green-to-tan body with black spots and either a red or a black head

Range: Throughout the United States and southern Canada

The fall webworm is a small pest moth found across the United States and southern Canada. In more northern locations, the adults are almost pure white, while those farther south tend to be more heavily spotted with brown, giving them an overall mottled appearance. However, it is the larvae that get all the attention. Able to feed on close to 100 different trees and shrubs, the developing larvae construct large airy webs by using silk to enclose the leaves at the end of one of the host's branches. The conspicuous webs can be several feet across, with infested trees often having multiple webs. The larvae feed together inside the web and expand it as they grow. Mature larvae are hairy and vary geographically, just like the adult moths. The primary difference is the color of the head: Individuals found in more northerly locations have red heads; farther south, their heads are black. While they can periodically defoliate host trees, the fall webworm is typically considered simply a nuisance.

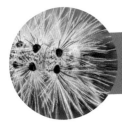

Believe It or Not: Although native to North America, the fall webworm was accidentally introduced to Europe and Asia where it is now a significant invasive pest.

Butterflies and Moths (Order Lepidoptera)

On Vegetation

Eastern Tent Caterpillar

Size: 1.5–2.0 inches long

ID Tips: Dark with a black head, long brown hairs, and light stripe down the back. Found in or near silken tents

Range: From the Rocky Mountains east in the United States

The eastern tent caterpillar is one of the most conspicuous early spring species. The young larvae begin to hatch from shiny black egg masses about the time that trees are leafing out. Gregarious through development, they start feeding together on the tender vegetation and construct the silken web or tent for which they are named. They use the tent, which may be seen on branches or even trunks, for protection and for temperature regulation, as it captures heat, much like a small greenhouse, a useful feature when springtime temperatures dip. Aided by the added heat, the larvae grow quickly and expand the tent to accommodate their increased size. When mature, the hairy caterpillars set out to find a location where they can spin their cocoons. While scouting out possible sights, they are often seen crawling across sidewalks and even on the siding of houses. A small brown moth eventually emerges, finds a mate, and lays eggs, which overwinter in preparation for the following spring. The adults are often found at artificial lights.

Believe It or Not: While typically just a cosmetic pest and often mistaken for the more problematic "armyworm," huge numbers can occasionally be produced during outbreak years, causing widespread tree defoliation.

Butterflies and Moths (Order Lepidoptera)

On Vegetation

Monarch Caterpillar

Size: 1.25–2.0 inches long

ID Tips: Body striped with alternating bands of black, white, and yellow; a pair of long, black filaments found on both ends

Range: Throughout the United States

Without a doubt, the monarch is one of the most iconic and commonly encountered garden caterpillars. Monarch caterpillars are specialist herbivores, feeding exclusively on plants in the milkweed family. The small conical cream-colored eggs are deposited singly on the flower buds, leaves, or stems of host plants. The resulting larvae are voracious eaters and can rapidly devour leaves. As they feed, the hungry caterpillars ingest toxins called cardiac glycosides, which render them and the eventual orange-and-black adult butterflies distasteful to many predators. The larvae advertise this chemical defense by having bold black, white, and yellow bands. While effective against many birds, other predators such as wasps, ants, and lizards readily attach and consume the plump caterpillars. Once fully grown, the caterpillar typically crawls off its host plant and seeks a secure place to pupate. Several generations may be produced each year. Attract monarchs by planting an assortment of colorful flowers and at least one species of milkweed. (When buying your plants, make sure they aren't treated with insecticides beforehand.)

Believe It or Not: Monarch caterpillars are known to respond to cool temperatures by developing more extensive black markings. The darker coloration helps them thermoregulate, which enhances their growth and development.

Butterflies and Moths (Order Lepidoptera)

On Vegetation

Black Swallowtail Caterpillar

Size: 1.8–2.2 inches long
ID Tips: Green with black bands and yellow-orange spots
Range: The eastern United States

The black swallowtail is one of the most commonly encountered garden butterflies, particularly east of the Rocky Mountains. While the sizable black-and-yellow adults are showy, their plump, green larvae tend to steal the show. Often referred to as "parsley worms," they feed on many cultivated herbs in the carrot family, including dill, fennel, and, of course, parsley. If disturbed, they rear back and extend a fleshy, orange hornlike structure called an osmeterium from behind their heads. This defensive gland releases a pungent odor and a chemical irritant that is particularly effective against predatory or parasitic insects, such as ants or wasps. When young, the larvae are mottled with black and resemble bird droppings. This unique disguise likely also helps them avoid being eaten. The black swallowtail is equally at home in undisturbed wetlands and rural meadows as it is in suburban yards and urban parks. Even a small container garden can support several caterpillars.

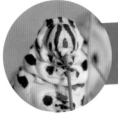

Believe It or Not: Female black swallowtails mimic the toxic pipevine swallowtail for protection from predators.

Butterflies and Moths (Order Lepidoptera)

Common Wood Nymph

Size: Wingspan 1.8–2.8 inches

ID Tips: Variable; broad brown wings with two large yellow-rimmed eyespots on the forewing; some populations have yellow patches surrounding the eyespots

Range: Throughout the United States and southern Canada

Found throughout the United States and southern Canada, this is one of our largest and most distinctive wood nymph butterflies. Adults are readily encountered in open grassy meadows and fields, although they may also frequent forest clearings and margins. They have a slow, relaxed flight and bob erratically along vegetation, stopping frequently to land within grasses or on the trunks of trees. Opportunistic feeders that frequently visit flowers, they are also at home near sap flows or fermenting fruit. The common wood nymph is geographically quite variable. In the eastern United States, most individuals have wide yellow patches on their forewings. By contrast, adults found throughout the West and in more northern locations lack these markings and are mostly a uniform brown. In total, there are 13 different recognized subspecies. The all-green larvae feed on grasses.

Believe It or Not: Like many other butterflies, the large eyespots of the common wood nymph likely serve a defensive purpose, deflecting the attack of a predator away from the insect's vulnerable body.

Butterflies and Moths (Order Lepidoptera)

On Vegetation

Promethea Silkmoth Caterpillar

Size: 1.85–2.4 inches long

ID Tips: Large, whitish-green with rows of small black spots; four red knobs on the thorax and a single yellow knob on the rear end

Range: The eastern United States

Like other giant silkmoth larvae, those of the promethea silkmoth are plump and quite showy. In fact, many people encounter the hungry caterpillars more often than the adult moths. After hatching from small egg clusters, the young larvae, now striped black-and-white, feed together on host leaves. They voraciously consume vegetation and grow quickly, molting their skin numerous times, which helps them adjust to their new size. They become more solitary with age and soon reach some two or more inches in length. Once mature, they begin to spin a cocoon by rolling up an individual leaf with silk. They make sure it stays on the plant through winter by reinforcing the stem with silk. The dangling elongated cocoons are light brown and easy to spot on bare branches. One generation is produced in northern sites, and up to two occur farther south.

Believe It or Not: Like many moths and butterflies, the developing larvae are sensitive to environmental signals. For example, promethea caterpillars monitor day length. Shorter days indicate that winter is coming and promote diapause (suspended development) until the following spring.

Butterflies and Moths (Order Lepidoptera)

Elegant Sheep Moth

Size: Wingspan 2.25–2.7 inches

ID Tips: Variable; forewings orange to pink with heavy black markings and a central black eyespot; hind wings paler yellow with heavy black markings and a central black eyespot

Range: The western United States and southern Canada

Aptly named, the elegant sheep moth is a large, boldly colored day-flying moth. Because of their size and bright pattern, adults are often mistaken for butterflies. They are quite common in open woodlands and shrubby habitats and are found from the Rocky Mountains west to California and north into southern Canada. While the pinkish-orange coloration of the adults is quite distinctive, the amount of black markings on the wings can vary tremendously. Female moths lay their eggs in rings around host branches. They overwinter and hatch the following spring. The spiny larvae feed together on the leaves when young and become solitary with age. While one generation per year is the norm, it may take them up to two years to complete development at higher elevations and more northerly locations.

Believe It or Not: Some spiders are known to mimic the sex pheromone of these moths, effectively luring unsuspecting males into their webs where they become a sizable and hearty meal.

Butterflies and Moths (Order Lepidoptera)

On Vegetation

Question Mark

Size: Wingspan 2.25–3.0 inches

ID Tips: Wings above orange with black spots; wings below brown with a dead leaf pattern; forewing tip squared off; hind wing has a stubby tail

Range: From the Rocky Mountains east

This handsome species gets its unique name from the small silvery hind wing spots that resemble (with some imagination) a rudimentary question mark. In sharp contrast to its bright orange upper surface, the wings below are cryptically mottled with brown, helping the butterfly resemble a dead leaf or tree bark when at rest. Found primarily in and near forests, the question mark is often spotted in wooded yards and neighborhoods. Adults have a strong, rapid flight but frequently alight on overhanging branches, tree trunks, or leaf litter. Wary and nervous, they are often difficult to closely approach. Both sexes feed at rotting fruit, animal dung, carrion, and tree sap, but not flowers.

Believe It or Not: The question mark produces distinct seasonal forms. Summer-form adults have mostly black hind wings; winter-form adults are predominantly orange.

Butterflies and Moths (Order Lepidoptera)

Viceroy

Size: Wingspan 2.6–3.2 inches
ID Tips: Wings orange with black veins and borders
Range: Throughout most of the United States except the western coast

The colorful viceroy seldom strays far from wetland habitats or other moist sites that support willows, its primary larval host. Males perch on overhanging vegetation and occasionally dart out to investigate passing insects or to take short exploratory flights. Although once thought to be a palatable mimic of the toxic monarch butterfly, studies have shown that both species are actually toxic to many predators. While the two butterflies are quite similar in appearance, the viceroy is somewhat smaller and has a distinctive black band through the middle of each hind wing. Adults are often encountered at a wide variety of flowers but will also feed at animal dung, carrion, and fermenting fruit.

Believe It or Not: In the extreme southern portion of its range, the vicerory is a deep mahogany in color and doesn't resemble the lighter-orange monarch as much.

Butterflies and Moths (Order Lepidoptera)

Red-spotted Purple

Size: Wingspan 3.0–3.6 inches

ID Tips: Wings above black with iridescent blue scaling on the hind wing; hind wing below brownish-black with a row of prominent orange spots

Range: The eastern United States and into the Southwest

This large eastern butterfly is one of several species that mimics the toxic pipevine swallowtail in order to scare off predators. It is commonly encountered in open woodlands, along forest edges, or in adjacent semi-open areas with young trees. It may also be a regular denizen of more forested suburban yards and parks. The colorful adults have a strong but gliding flight and regularly perch on sunlit branches. They occasionally visit flowers but often prefer rotting fruit, animal dung, carrion, or tree sap. In areas where their ranges overlap, they readily hybridize with the more northern white admiral butterfly.

Believe It or Not: Females deposit their eggs singly on the tips of host leaves. The young larvae begin feeding, constructing a long chain of fecal pellets and debris on which to perch. This unique platform presumably helps them avoid detection from predators.

Butterflies and Moths (Order Lepidoptera)

Tomato Hornworm

Size: 3.0–4.0 inches long

ID Tips: Bright green with seven white V-shaped marks on the side of the body; a prominent curved horn off the rear end

Range: Throughout most of the United States

This impressive caterpillar is about the size of a hot dog when fully grown. Along with the tobacco hornworm, its similar-looking relative, it is regularly encountered in vegetable gardens where it feeds voraciously on tomato plants. Although large, their green coloration helps them blend in quite well with surrounding vegetation. As a result, most gardeners tend to notice the feeding damage before ever spotting the culprits. Just one or two larvae can quickly defoliate a sizable plant. They may feed on the leaves, blossoms, or even developing tomatoes. Once feeding is complete, the mature larvae crawl down the stem and burrow in the soil to pupate underground. The resulting adult moths are a relatively drab gray with elongated barklike forewings and noticeable yellow spots on the abdomen. Impressive in size, their wingspan may approach five inches.

Believe It or Not: Adult moths have a proboscis up to 4.5 inches long or more. This allows them to feed at a wide variety of long, tubular blossoms that are inaccessible to many other insects.

Butterflies and Moths (Order Lepidoptera)

On Vegetation

Cecropia Moth Caterpillar

Size: 4.0–4.5 inches long

ID Tips: Large; bluish-green body with bright blue, yellow, and red tubercles, each with short, black spines

Range: The eastern United States

This impressive insect is one of the most beautiful North American moths. Adults have colorful black-and-orange patterned wings and a fuzzy red-and-white striped body. Like most other giant silkworm moths, they are active by night and periodically attracted to artificial lights. Despite its striking appearance, the cecropia moth may be overshadowed by its equally eye-catching, sizable, and more frequently encountered caterpillar. Approaching some five inches in length when fully grown, it is gaudily adorned with an assortment of bright colored spiny knobs. The growing larvae have ferocious appetites and quickly devour the leaves of a wide assortment of host trees and shrubs, including maple, willow, and birch, just to name a few. When fully grown, they spin a large papery silken cocoon that is tapered at both ends and firmly attached horizontally to a small twig or branch.

Believe It or Not: The cecropia moth and other giant silkworm moths do not have a functioning mouth and don't feed as adults. Instead, they rely on the food reserves acquired during their larval stage to survive their short adult lives.

Butterflies and Moths (Order Lepidoptera)

On Vegetation

Regal Moth/Hickory Horned Devil

Size: 4.5–5.5 inches long

ID Tips: Large, greenish-blue body with an orange head; long, curved red-and-black horns

Range: Primarily the southeastern United States; into the Northeast and the Midwest

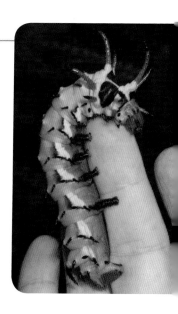

Although large and brightly colored, adult regal moths are frequently overshadowed by their imposing caterpillars. Aptly named hickory horned devils, full-grown caterpillars may exceed five inches in length and are as plump as a bratwurst. Resembling small dragons, the plump, blue-green caterpillars are armed with long red-and-black curved horns on the thorax. Despite this ferocious appearance, they are completely harmless. They feed on the leaves of various trees and shrubs, including hickory, walnut, and sumac. Hickory horned devils do not spin cocoons, pupating in the soil instead. As a result, they are frequently encountered on the ground when they are searching for a suitable burrowing site.

Believe It or Not: The hickory horned devil is the largest caterpillar in North America.

Grasshoppers, Crickets, Locusts, and Others (Order Orthoptera)

On Vegetation

Snowy Tree Cricket

Size: 0.5–0.75 inch long

ID Tips: Pale green body with long antennae and transparent wings; male has broad wings that fold over the back, producing a flattened teardrop-like appearance; females have wings that curl around the body, producing a narrower, more streamlined shape

Range: Most of the United States except the Deep South

As their name implies, snowy tree crickets spend their lives within the branches of trees and shrubs, preferring plants with broad leaves. Inactive during the day, they rest quietly, often in protected areas, to avoid being discovered by predators. But come nightfall, they actively forage for food among vegetation, feeding both on leaves and other small insects. They have a particular preference for aphids, especially young ones. Adult snowy tree crickets are best known for their song, a series of melodious chirps that males produce in order to attract potential mates. Synchronous choruses of many males occur and can be extensive.

Believe it or not: The rate of the call is influenced by temperature. As a result, you can use the cricket as a natural thermometer.

Grasshoppers, Crickets, Locusts, and Others (Order Orthoptera)

Katydid

Size: 0.5–2.5 inches long

ID Tips: Generally green body with large hind legs and very long thread-like antennae; may have large wings; females have a prominent curved ovipositor that's upright and projects from the end of the abdomen

Range: Throughout the United States

Although there are more than 250 katydid species in the United States, a great many resemble large green crickets at first glance. They are primarily nocturnal and spend the majority of their lives in vegetation, often high in trees, munching away on available leaves. Most rely on camouflage for protection, blending into the surrounding greenery. While many katydids have large wings, they tend to be poor fliers, relying instead on gliding or simply walking slowly. They are probably best known though for their distinctive raspy courtship calls, which are produced primarily by males who rub both forewings together in rapid succession. They get their odd name from the distinctive song of the common true katydid. As males start to call, neighboring individuals often chime in but in alternate fashion, resulting in an amusing verbal quarrel that sounds like "katy-did, katy-didn't."

Believe It or Not: Katydids hear with structures on their front legs that can convert airborne sound waves to fluid-borne sound waves, much like the human ear.

Grasshoppers, Crickets, Locusts, and Others (Order Orthoptera)

On Vegetation

Red-legged Grasshopper

Size: 0.75–1.50 inches long
ID Tips: Elongated yellow body with a brown back; bright red hind legs
Range: Much of North America

Found across much of North America, this is a common grasshopper of open, weedy areas, including roadsides, pastures, grasslands, agricultural fields, and yards. The adults and nymphs are active during the day and feed on a wide range of different plants, periodically becoming pests of several crop species. At night, the insects rest high on weedy vegetation and grasses. Population numbers tend to fluctuate considerably from year to year. Females lay several clusters of eggs in the soil, which overwinter and hatch the following spring. Depending on their available diet, a single female can produce several hundred eggs in her lifetime.

Believe It or Not: Adults often have a range that varies with the weather; in times of drought, adult red-legged grasshoppers often develop longer wings and are more active fliers, often traveling extensive distances in search of nutritious food.

Grasshoppers, Crickets, Locusts, and Others (Order Orthoptera)

On Vegetation

Differential Grasshopper

Size: 1.0–2.0 inches long

ID Tips: Elongated yellow-brown to olive-greenish; black crescent marks on the hind legs

Range: Throughout most of the United States

This drab-colored insect is one of the most common grasshoppers in the United States. Like other grasshoppers, the adults and nymphs are herbivores and feed on a broad range of different plants, readily moving from one food source to another, depending on quality and availability. It is considered an agricultural pest of numerous crops, including everything from corn to fruit trees, and can cause significant damage or even complete defoliation in outbreak years. Large swarms have been known to destroy entire crop fields in just a matter of days. More often, though, the differential grasshopper is encountered in a variety of open, grassy, or weedy areas, including old fields, pastures, gardens, and roadsides. Generally only a single generation is produced each year.

Believe It or Not: Adults are very strong fliers and are capable of traveling many miles. They may also rise up on wind currents and have been reported to periodically reach heights of 1,000 feet or more.

Grasshoppers, Crickets, Locusts, and Others (Order Orthoptera)

On Vegetation

Carolina Grasshopper

Size: 1.2–2.0 inches long
ID Tips: Mottled grayish to brown; body is elongated
Range: Most of the United States and southern Canada

Also called the Carolina locust, this large grasshopper is common across much of the eastern United States and southern Canada. It prefers dry, open habitats with bare patches and can be commonly found in fields, abandoned lots, along railroad tracks, or on gravel and dirt roads. Both adults and nymphs are drab colored and highly camouflaged, enabling them to blend in with the bare earth or stones. However, upon close approach, adults readily fly up to reveal their conspicuous cream-bordered black hind wings, and they may make a noticeable flapping noise in the process. At first glance, a flying Carolina grasshopper may resemble a large butterfly. But the adults quickly land again on the ground several yards ahead. The species feeds on a number of different plants and grasses and can be a minor crop pest in years when they are abundant.

Believe It or Not: Adult Carolina grasshoppers are strong fliers and may occasionally disperse up to several miles in search of available food.

Grasshoppers, Crickets, Locusts, and Others (Order Orthoptera)

On Vegetation

Eastern Bird Grasshopper

Size: 1.5–2.75 inches long
ID Tips: Light brown to reddish-brown body with light and dark bands and dark, spotted wings
Range: The eastern United States

This large insect is also called the American grasshopper. Common throughout most of the Eastern United States, it is large and commonly found in fields and forest margins. It feeds on a wide range of plants and grasses, including many crop and ornamental species and can periodically cause significant feeding damage, especially when present in high numbers. They are strong fliers and can disperse over long distances in search of food. Unlike many grasshoppers, they tend to be somewhat arboreal, often feeding and resting in trees and other tall vegetation. In the southern portions of its range, it overwinters as an adult and produces two annual generations. Females insert their abdomens in the soil and lay their eggs in clusters underground. The newly hatched nymphs dig their way to the surface and begin feeding together in large groups, becoming more solitary as they develop.

Believe It or Not: American bird grasshoppers can occasionally occur in abundance and can form large swarms. The resulting feeding damage can be severe and may result in complete defoliation of affected vegetation.

Grasshoppers, Crickets, Locusts, and Others (Order Orthoptera)

On Vegetation

Eastern Lubber Grasshopper

Size: 2.25–3.1 inches long

ID Tips: Large; robust dull yellow with black spots and markings; short wings with a pinkish hue

Range: The southeastern United States

This insect is impressive in both size and color. Ranging throughout the Southeast, it cannot be confused with any other grasshopper in the region. Their bright colors advertise to potential predators that they are chemically protected. If molested, eastern lubber grasshoppers may also emit an acrid secretion from their abdomens. Nymphs often congregate on vegetation, further emphasizing the warning to predators. At home in open habitats such as weedy fields and pine forests, they can be quite abundant and are regularly encountered. Unable to fly, their short, stubby wings barely make it halfway down their abdomens. They instead rely on hopping and crawling to move, and are often quite clumsy in doing so. Lubbers are highly variable, changing color as they develop and also having distinct geographic forms.

Believe It or Not: Eastern lubber grasshoppers are easy to keep in captivity. As a result, they are commonly used for education and display by zoos, schools, and insectariums.

Spiders (Order Araneae)

On Vegetation

Jumping Spider

Size: 0.15–0.6 inch long
ID Tips: Generally small; stout, hairy body with two of its eight eyes very large and forward-looking
Range: Throughout the United States

Jumping spiders are an extremely diverse and species-rich group. Despite their tiny size and somewhat adorable appearance, they are ferocious hunters. They actively search for food during the daytime, feeding on a wide range of other insects, including many organisms much larger than themselves. Jumping spiders have excellent vision, using their well-developed eyes to detect motion, and they can judge spatial relationships very accurately. This allows them to stalk potential prey and ultimately pounce on them from some distance, often leaping many times their body length. When doing so, jumping spiders release a silken safety line behind them, just in case. This allows them to retreat back to their perch if they are unsuccessful. They are common in most yards, as well as adjacent weedy areas, and may often be found in homes. Look for them on windowsills.

Believe It or Not: Jumping spiders rely on blood pressure—not strong muscles—to leap. They can rapidly increase the blood flow to their legs (blood is known as hemolymph in arthropods), causing the spiders to extend suddenly, propelling the them forward.

Spiders (Order Araneae)

On Vegetation

Black and Yellow Garden Spider

Size: 0.75–2.5 inches long
ID Tips: Abdomen egg-shaped with black-and-yellow markings; eight black legs marked with yellow or red; females are much larger than males
Range: Throughout the United States

Also called the common garden spider, this is one of the largest, most conspicuous, and abundant orb-weaving spiders in North America. Adults prefer to construct their impressive circular webs, which may measure over two feet in diameter, in open sunny locations, including yards, gardens, fields, and forest edges. Each web is an ornate structure anchored to adjacent vegetation and marked with a distinctive central zigzag pattern. Adults rest head-down in the web and await unfortunate victims that venture too close. Once ensnared in the sticky strands, there is no escape. The vibration caused by the struggling insect alerts the spider, which quickly wraps the prey in silk and disables it with paralyzing venom. After mating, each female constructs several small, round papery brown egg sacs that may each contain more than 1,000 eggs. The resulting young spiders spend the winter inside the protective, insulated sac, becoming active and dispersing the following spring. Although fearsome-looking, black and yellow garden spiders are harmless to humans and are a welcome beneficial predator of many pests.

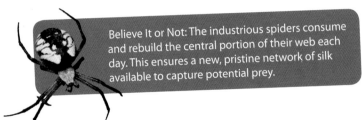

Believe It or Not: The industrious spiders consume and rebuild the central portion of their web each day. This ensures a new, pristine network of silk available to capture potential prey.

True Flies (Diptera)

Long-legged Fly

Size: 0.05–0.35 inch long

ID Tips: Small, often with a metallic body; bright eyes; mostly transparent wings and long, thin legs

Range: Throughout the United States

These insects are easily overlooked due to their small size. While found in a wide variety of different habitats, long-legged flies are often seen perched on leaves along partially shaded woodland margins. In direct sunlight, their bright metallic bodies shine in shades of blue, green, and copper and rest above noticeably elongated, thin legs. Many also have distinctly bright shiny eyes that may often appear red. They are easily disturbed and often nervously fly up, only to land a few seconds later on the same or nearby leaf. The adults are predatory and feed on an assortment of other small insects. The larvae of most species are also predaceous and develop in soil, decaying organic material, or under bark where they hunt small invertebrates.

Believe It or Not: Many male long-legged flies perform intricate courtship displays and have modified legs or antennae to enhance the display.

True Flies (Diptera)

On Vegetation

Robber Fly

Size: Variable, typically 0.3–1.0 inch long

ID Tips: Highly variable in appearance; typically slender body with a tapered abdomen; large eyes; and long, bristly legs

Range: Throughout the United States

Like miniature hawks, robber flies are swift-winged predators with large eyes and good vision. Common in open sunny areas, adults perch on leaves, twigs, or other prominent features and fly out to snag passing insects with their long, barbed legs. They quickly return to the perch and inject their prey with saliva that contains a neurotoxin and enzymes. The neurotoxin paralyzes the insect and the enzymes break down the body contents into a liquid that can be easily consumed. Robber flies are incredibly diverse. They come in a great many shapes and sizes, with some even resembling bees or wasps. Nearly 1,000 species can be found in North America alone. The wormlike robber fly larvae are also predatory, living in soil or decaying material where they feed on small, soft-bodied insects.

Believe It or Not: Robber flies are members of a large and diverse family with more than 7,000 species found worldwide.

Dragonflies and Damselflies (Odonata)

On Vegetation

Ebony Jewelwing

Size: 1.5–2.0 inches long
ID Tips: Large; dark body with broad, dull, black wings
Range: From the Rocky Mountains east

The ebony jewelwing is one of the largest, most distinctive, and truly spectacular damselflies in eastern North America. Males have brilliant, metallic blue-green bodies with all-black wings. Females are a bit duller and have a prominent white spot on the tip of their wings. Adults have a leisurely, somewhat fluttering flight and pause frequently to perch on sunlit vegetation, where they are often easy to closely approach. If disturbed, the adults typically fly a short distance before alighting again. The ebony jewelwing is common along slow-moving streams or rivers in forested areas but may venture quite far into nearby open habitats. Unlike their more robust dragonfly cousins, damselflies have elongated, delicate bodies and hold their wings together over their backs when at rest.

Believe It or Not: Male ebony jewelwings regularly establish territories along streams and actively guard egg-laying females, defending them from the advances of rival males.

Mantises (Order Mantodea)

On Vegetation

Praying Mantis

Size: Variable, 2.0–4.0 inches long

ID Tips: Long, slender green to brown body; triangular head; enlarged front legs for grasping prey

Range: Throughout the United States

Distinctive and engaging, the praying mantis is truly a remarkable insect. Masters of camouflage, they sit motionlessly on plants, waiting to ambush unsuspecting organisms that wander too close. Once in range, the prey stands little chance. In the blink of an eye, the mantis reaches out with its powerful front legs, which are armed with spikes to ensure a good grip, and dinner is served. Depending on its size and developmental stage, a mantis will eat a variety of insects, including its brothers and sisters, if the opportunity arises. They will even eat small lizards or frogs. They have a characteristic triangular-shaped head, large eyes, and binocular vision to scan the landscape for food. This is also the only insect that can look over its shoulder, turning its head up to 180 degrees. The developing nymphs are smaller versions of the adults, but with black wings. Late in the season, mature females produce a Styrofoam-like sac containing many eggs. The eggs overwinter and hatch the following spring.

Believe It or Not: A praying mantis can make a great pet. They are long-lived, intelligent creatures that can be easily maintained in a small cage or aquarium. Remember, though, that they are living creatures and require regular care.

Walking Sticks (Order Phasmida)

On Vegetation

Walking Stick

Size: 3.0–3.6 inches long

ID Tips: Long, slender brown or green body; long legs and antennae; resembles a small twig or branch.

Range: Throughout the United States

As their name suggests, walking sticks are true masters of concealment. They literally take camouflage to the next level and seamlessly blend in to their environment by looking like something they are not. Both adults and immature specimens have extremely long legs and slender bodies that resemble twigs or small branches. If discovered and under threat, they have another trick at their disposal. Walking sticks can discharge a foul-smelling chemical, spraying it with very accurate aim. The resulting surprise secretion is quite effective against many predators. These distinctive insects are herbivores and feed on a variety of different plants, staying safely in the vegetation their entire lives. Walking sticks are easy to care for and can make good summer pets.

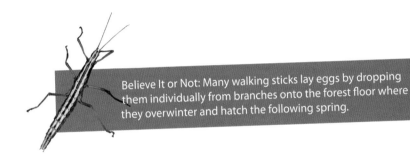

Believe It or Not: Many walking sticks lay eggs by dropping them individually from branches onto the forest floor where they overwinter and hatch the following spring.

Fun and Family-Friendly Bug Activities

These fun, simple projects and activities are designed to help you discover, observe, and learn more about the insects in your area. They are inexpensive, and offer a great opportunity for the whole family to participate. The following activities are ranked from simplest to most advanced. **Note:** Children should always complete activities under the supervision of an adult.

Easy

Netting Insects

WHAT YOU'LL NEED
☐ An insect net ☐ A jar with a lid

One way to safely and easily look at insects is with the use of a butterfly net. Simple high-quality insect nets are available from many entomological and scientific supply companies. A net allows you to catch unknown or interesting insects at flowers, on the ground, or on vegetation. Once in the net, you can use a large clear plastic jar with a lid, such as an old peanut butter jar, to remove the organism. Safely in the jar, the insect can be closely inspected without worry of injuring it or having it bite or sting you. When finished, simply open the lid and let the insect out. Be careful, however, as some stinging insects can be aggressive. If it's a stinging insect, or you think it might be, remove the lid and quickly fling the jar while holding on to it. This will help catapult the organism out, ensuring a safer release distance between you and any danger.

Insect nets can also be used to sweep through tall grasses and vegetation or along the branches of trees or shrubs. If you sweep quickly with the net, you can collect many organisms that you might otherwise not have noticed. Use the same plastic jar to remove the organisms and inspect them. Be sure to release all the captured insects back in the same general vicinity when you are finished.

Hunting for Wolf Spiders with a Flashlight
Wolf spiders are like small tarantulas. Both are sit-and-wait predators and active hunters, and they tend to be most energetic at night. A wolf spider's eyes reflect light directly back to its source, and this makes one easy to spot with a simple handheld flashlight. Simply shine the beam from a flashlight along the ground and amid leaf litter in wooded areas, and then just look for the small glowing eyes staring back at you.

Butterfly Watching
Butterflies are fun and entertaining insects to observe. They are commonly encountered in yards, gardens, parks, old fields, and nearby wild areas, especially those with many blooming flowers.

WHAT YOU'LL NEED

☐ A butterfly field guide appropriate to your state or region.

☐ A digital camera or a smart phone to take photos of what you find. Butterflies are often challenging to identify at first sight. Having a picture to review can be very helpful, as it helps you compare them to a field guide or website.

☐ A notebook and a pen to record the species you see and any specific notes. Be sure to document the date and location of your observation, the time of day, the sex of the butterfly, if you can identify it, and any interesting behaviors or other observations. If you record your sightings, you can contribute valuable data to citizen science programs in the process. Helpful online resources or identification and observation recording include Butterflies and Moths of North America (www.butterfliesandmoths.org), BugGuide (www.bugguide.net), iNaturalist (www.inaturalist.org), and e-butterfly (www.e-butterfly.org).

Attracting Insects with a Black Light
Many insects are active at night and can easily be attracted to artificial lights. Entomologists generally use black lights, or ultraviolet (UV) lights, as these are more effective than traditional incandescent lights at drawing in a wider array of insects. Black lights are relatively inexpensive and can be purchased from many entomological or scientific supply companies. They can run on a portable battery, car battery, or via an electrical outlet using an extension cord,

whichever is easiest. If you can't afford an entomological black light, you can purchase a UV blacklight flashlight for around $10 at many department stores or online. If neither is an option, a standard incandescent light can be used as a substitute, but it will generally be less effective.

WHAT YOU'LL NEED
- ☐ An old white sheet
- ☐ A good length of rope or twine
- ☐ 2 binder clips
- ☐ 1 blacklight
- ☐ 2 small rocks or bricks

Locate two trees and tie the end of the rope or twine to one tree about 5 to 6 feet above the ground. The two trees should be at least the width of the white sheet apart. Run the rope or twine to the other tree, and loop the rope around the tree at least once, ensuring the rope is level and taunt. Then run the rope back to the original tree and tie it. When you're done, you'll have essentially created a clothesline. Hang the white sheet from one rope so that the bottom of the sheet is lying on the ground. Secure the top of the sheet to the rope with the binder clips. Clothespins also work well. Pull the bottom of the sheet forward a bit, creating a 90-degree angle with the ground, and place a rock or brick on each bottom corner. This will help keep the sheet from flapping in the wind and allow you to observe any insects that fall off the sheet or crawl onto it from the ground. Suspend the light from the other rope so that it hangs down in front of the sheet from about halfway up. Turn the light on and wait for the insects to arrive. Ideally the sheet and light should be set up and powered on by sunset. For best success, choose dark, cloudy nights and avoid times with a full moon or a nearly full moon. Also, choose locations away from other competing artificial lights.

Glow-in-the-Dark Insects and Arthropods with a Handheld Black Light

Many organisms glow under ultraviolet light (a blacklight). The list includes scorpions, centipedes, and spiders, as well as many caterpillars and other insects. This makes a handheld blacklight (an ultraviolet light) the perfect tool to explore the nocturnal world. You can buy one online or in many large retail stores. They are battery operated, safe to use, and easy to carry. Once you have one, head outdoors at night and use the UV light to explore, moving it slowly over leaf litter, vegetation, logs, or other hidden places and see what you might discover.

More Advanced Projects

These projects require a bit more know-how or supplies, but they're well worth pursuing.

Moth Baiting

Many moths (and some other insects) can be attracted to bait at night. This technique is often called sugaring or baiting. There are many specific bait recipes available online, and each differs slightly in terms of ingredients. A simple and effective one can be found at nationalmothweek.org/finding-moths-2/daves-recipe-for-moth-bait/

WHAT YOU'LL NEED

☐ 1 wide-mouth plastic container with lid   ☐ 1 paintbrush

☐ Bait ingredients. The specific recipes differ but usually include fruit (bananas, watermelon, peaches, apples), sugar (or molasses, maple syrup, or honey), and alcohol (dark rum and/or beer).

Once you've made your bait, apply the somewhat strong-smelling, fermented mixture of fruit, sugar, and alcohol to the trunk of a tree, a fence post, a boulder, a log, a stump, or another sizable structure prior to sunset. Then check the baited area regularly through the night, and see which organisms have arrived to feed on the bait.

Native Bee Nest Box

Most native bees are solitary and non-aggressive. They are highly beneficial to the environment, fun to observe, and easy to attract. Bees depend on flowers for food, but they also need a place to nest. While many native bees nest in the ground, others prefer to nest in small tunnels, such as hollow stems or the holes left behind from wood-boring beetles. These tunnel-nesting species are easy to attract with artificial nest boxes. Such nest boxes can also help boost local bee populations. Commercial nest boxes are widely available, and you can purchase from many suppliers, including garden centers, nature stores, and various online retailers. However, they are quite easy to construct on your own, and they require little investment of time or money. If you want to make your own, there are two options. You can make a wooden block nest or construct one out of stem bundles.

WHAT YOU'LL NEED

☐ 1 or more untreated blocks of natural, untreated wood; the ideal dimensions are around 4 inches wide, 6 inches deep, and anywhere between 6 and 16 inches long. (If large blocks of wood are unavailable, then several pieces of lumber can be glued together to achieve the appropriate dimensions.)

☐ A drill and multiple drill bits (¼" to about ⅜" are ideal)

☐ A bracket for mounting or hanging the box

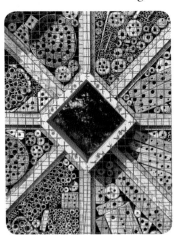

Place the wooden block on the ground or on a workbench. Using the various diameter bits, drill into the block on only one side starting about 1" from each edge. Be sure to drill straight into the block and avoid angled holes. Aim for a hole depth of 3 inches or more, but be careful to not drill completely through the block. Alternate diameter drill bits as you go, spacing them about 1" apart until the block is filled with holes of varying sizes. Secure the bracket on the top or back, and mount firmly on a building, a fence post, a tree, or another structure in a protected location with access to the morning sun. (The same technique can be used to drill various diameter holes into a dead stump.) Bees will then build nests in the holes, where you can observe them from a distance.

WHAT YOU NEED FOR A STEM BUNDLE NEST

☐ A bundle of reeds, bamboo, hollow stems, and/or paper straws, all of the same length. If you use bamboo or reed, be sure to cut it so that one end of the plant is open, and one is closed.

☐ Wire or a zip tie

Take a handful of the variously sized materials and strap them together in a very tight bundle using the wire or plastic zip tie. Depending on size, one or more bundles can then be placed, open-end out in a container. These can vary from a wood-frame box or a hollowed-out branch to a piece of PVC pipe with a cap at one end. The container with the nest box should be placed in a sheltered location with morning sun, several feet off the ground, and with the bundles perpendicular to the ground. There are a wide range of potential designs. More detailed instructions are readily available online from many sources.

Planting a Pollinator Garden

A wide range of insects feed on the nectar and pollen produced by flowering plants. Plants with colorful blossoms help draw these organisms into a particular location, often significantly concentrating their numbers on a few productive plants. The same flowers also often attract various predators, such as spiders, praying mantises, and ambush bugs, which feed on the bounty of available prey insects. Thus, planting a pollinator garden is one of the easiest ways to observe many local insects and spiders. Even a small area or container planted with flowers can attract a great variety of species to your yard.

Tips for Gardening for Bugs

1. Select a variety of flowers with different colors, shapes, and sizes. This will help attract the widest variety of insects. You can use both annuals and perennials.

2. Aim to have plants that bloom throughout the growing season. Many insects are active well into late fall and also early in spring.

3. Do your homework. Understand the light, soil, and water requirements of each plant before buying it. When in doubt, ask a nursery professional or look the plant up online. This will ensure that you include the right plants for your specific geographic area and growing conditions.

4. After planting, be sure to regularly water your plants until they are firmly established.

5. Watch and enjoy the insects and other organisms that visit your flowers. You may wish to purchase a pair of close focus binoculars to help with observation.

6. Avoid using pesticides or herbicides in your yard, as these can harm insect populations.

Rearing Caterpillars

Butterfly and moth caterpillars are commonly encountered in most yards and neighboring wild spaces. Taking care of caterpillars requires time and commitment, but it can be both fun and rewarding. Doing so gives you an opportunity to enjoy them, watch them grow, pupate, spin a cocoon and eventually see an adult butterfly or moth emerge. Caterpillars have relatively simple needs. They require a regular supply of fresh, high-quality plant material; clean conditions; and a safe, secure, and roomy environment in which to be housed. Various containers can be used, depending on the size and number of caterpillars. While plastic cups or other containers are useful, often a mesh cage or an aquarium with a lid are the best options. Avoid tightly sealed containers, as this can cause humidity or moisture to build up, which can lead to mold and diseases.

In general, only rear caterpillars that you find feeding on vegetation. This will ensure that you know and have access to the correct food resources. You can use either potted plants or cut plant material. If you're using cut plant material, place the stems into a plastic container of water; old plastic soda, water, or other beverage bottles work well. Place the container with leaves in the cage or aquarium and place the caterpillars on the leaves. Make sure the container holding the water doesn't have a large gap; otherwise, the caterpillar could crawl down into the water and drown.

Then cover the aquarium or cage in order to avoid any caterpillars from escaping. Replace the leaves and clean out the cage or aquarium daily. Some caterpillars may take weeks to fully develop. Remember that the caterpillar will die without healthy food or if conditions are unclean. Once the caterpillar is fully grown, it will typically crawl off the leaves and wander to find a secure place to pupate or spin a cocoon. Once it has done so, place a few extra sticks or branches in the cage so that the eventual adult butterfly or moth has something to crawl up on and expand its wings. Cocoons or pupae produced in late summer or early fall may overwinter, with the adults emerging in spring. If it is late in the season and the adults have not emerged after about 3 to 4 weeks, you can place them in a cool, dark location such as a garage or basement. This should help keep them from inadvertently emerging until the following spring. Then, when the weather warms, bring them back into the house or onto a porch in a bright location out of direct sunlight. This should encourage the adult moth or butterfly to emerge, and you can observe it for a short time before releasing it into your yard or a nearby wild area.

Pitfall Trapping

Pitfall traps are frequently used by entomologists to sample ground-dwelling insects and other arthropods. Assembling one is simple and easy; you can do so in just a short period of time using materials found around the home.

WHAT YOU'LL NEED

☐ A plastic jar with a wide lid. An old peanut butter jar or an 8 oz. drinking cup works great.

☐ A small garden trowel   ☐ 3 small rocks or stones of equal size

☐ A flat piece of wood or a tile that is larger than the jar/cup opening

Using the garden trowel, dig a hole in the ground large enough to fit the container. Its rim should be level with the soil surface. Backfill tightly around the container so that there are no gaps. Place the small stones at equal distances around the rim at least 1 inch away from the opening. Cover the container with the board or tile, ensuring that it is resting on the stones and raised above the opening. This will help keep rain out of your pitfall trap and ensure a safe, shady environment for any organism that happens to fall in. Check your trap regularly (about once every 24 hours is ideal). For best results, set up several traps around your yard to increase the chances of catching interesting insects.

About the Author

Jaret C. Daniels, Ph.D., is a professional nature photographer, author, native plant enthusiast, and entomologist at the University of Florida, specializing in insect ecology and conservation. He has authored numerous scientific papers, popular articles, and books on gardening, wildlife conservation, insects, and butterflies, including butterfly field guides for Florida, Georgia, the Carolinas, Ohio, and Michigan. He is also coauthor of *Wildflowers of Florida Field Guide* and *Wildflowers of the Southeast Field Guide*. Jaret currently lives in Gainesville, Florida, with his wife, Stephanie.